HYDROGEN ENTRY and ACTION in METALS

Pergamon Titles of Related Interest

International Association for Hydrogen Energy
DIRECTORY OF HYDROGEN ENERGY PRODUCTS AND SERVICES
International Association for Hydrogen Energy
HYDROGEN IN METALS
Veziroglu HYDROGEN ENERGY PROGRESS I
Veziroglu METAL-HYDROGEN SYSTEMS
Williams HYDROGEN POWER

Related Journals*

ACTA METALLURGICA
CANADIAN METALLURGICAL QUARTERLY
ELECTROCHIMICA ACTA
ION-SELECTIVE ELECTRODE REVIEWS
METALS FORUM
PROGRESS IN SURFACE SCIENCE

*Free specimen copies available upon request.

HYDROGEN ENTRY and ACTION in METALS

Malcolm A. Fullenwider

Pergamon Press
New York   Oxford   Toronto   Sydney   Paris   Frankfurt

Pergamon Press Offices:

U.S.A. Pergamon Press Inc., Maxwell House, Fairview Park, Elmsford, New York 10523, U.S.A.

U.K. Pergamon Press Ltd., Headington Hill Hall, Oxford OX3 0BW, England

CANADA Pergamon Press Canada Ltd., Suite 104, 150 Consumers Road, Willowdale, Ontario M2J 1P9, Canada

AUSTRALIA Pergamon Press (Aust.) Pty. Ltd., P.O. Box 544, Potts Point, NSW 2011, Australia

FRANCE Pergamon Press SARL, 24 rue des Ecoles, 75240 Paris, Cedex 05, France

FEDERAL REPUBLIC OF GERMANY Pergamon Press GmbH, Hammerweg 6, D-6242 Kronberg-Taunus, Federal Republic of Germany

Copyright © 1983 Pergamon Press Inc.

Library of Congress Cataloging in Publication Data

Fullenwider, Malcolm A. (Malcolm Allan), 1940-
  Hydrogen entry and action in metals.

  Includes index.
  1. Metals--Hydrogen content. 2. Metals--Hydrogen embrittlement. I. Title.
TN690.F78   1982   669'.94   82-12199
ISBN 0-08-027526-5

All Rights reserved. No part of this publication may be reproduced, stored in a retrieval system or transmitted in any form or by any means: electronic, electrostatic, magnetic tape, mechanical, photocopying, recording or otherwise, without permission in writing from the publishers.

Printed in the United States of America

Contents

PREFACE		vii
NOMENCLATURE		ix
CHAPTER		
1	Introduction	1
2	Diffusion Studies: The Bi-Electrode Technique	4
3	Diffusion Studies: Barnacle Electrode	26
4	Statistical Studies	53
5	The Hydrogen Electrode	62
6	Catalysis	78
7	Storage and Purification of Hydrogen as Metal Hydrides	88
8	Embrittlement	95
APPENDIX A:	Infinite surface source, equilibrium, diffusing into a semi-infinite solid	107
APPENDIX B:	Rediffusion from erfc initial condition	109
APPENDIX C:	Constant flux, diffusing into a semi-infinite solid	112
REFERENCES		115
INDEX		121
ABOUT THE AUTHOR		125

Preface

The purpose of this book is to arrange, in one publication, work on hydrogen in metals done by American researchers centered about electrochemical techniques. European work can be found described elsewhere in literature.

There is certainly much to be said for the care and conservatism which European scientists take in their work. The attitude that no large task can be taken on at a small university, such as most European universities, is much said and found implicitly in their work. In any case, the work presented here has been largely neglected by European authors.

The author was introduced to the field of hydrogen diffusion in metals while working at the Electrochemistry Laboratory of the University of Pennsylvania in the 1960s under Professor John O'M. Bockris. The author immediately found the project, the study of hydrogen diffusion in a variety of metals with the bi-electrode technique, very interesting, and many thanks go to John Bockris for proposing such a fascinating topic. It was during this period that everything was "dropped" for an interval to work on the "Barnacle Electrode," a device originally described in a Navy proposal by Drs. Leonard Nanis and Jim McBreen in 1962. This brief episode has recently resulted in a calibrated model of the electrode, the last efforts being due to Drs. John Deluccia, Dave Burman, and Vinod Agarwala of the Naval Air Engineering Center, Warminster.

I would like to thank the people who read the manuscript in its various forms of preparation: Professor E. Schmidt for his checking of the equations, Dr. P. K. Subramayan for his help from the start of the book, and Dr. Jim McBreen; and, of course, Professor John Bockris who helped from the very beginning of my involvement in the subject 20 years ago. And thanks go to the typist, Leslie Frable, for the care she took with the manuscript.

Also, I thank my family, Susan, Chris, and Mike, for keeping me cheered up during the course of this work.

Nomenclature

c = concentration, moles cm^{-3}

x = distance, cm

D = diffusion coefficient, $cm^2\ sec^{-1}$

t = time, sec

T = temperature, °C. and °K

κ = thermal diffusivity, $cm^2\ sec^{-1}$

c_o = maximum concentration just inside surface of diffused system, moles cm^{-3}

δ = thickness of diffused specimen, cm

$\bar{c}$ = Laplace transform of c

p = Laplace transform variable and pressure of hydrogen gas, atmospheres

q = $(p/D)^{1/2}$

J_t = flux, variable, moles $cm^{-2}\ sec^{-1}$

$\bar{J}$ = Laplace transform of J

a = constant

b = constant

$\hat{\Theta}_2$ = a theta function

J_∞ = steady state flux, moles $cm^{-2}\ sec^{-1}$

σ_H = hydrostatic stress

NOMENCLATURE

$C_{,\sigma H}$ = concentration of hydrogen at stress σ_H

$c_{H,o}$ = concentration of hydrogen at zero stress

$\bar{V}$ = partial molar volume

R = gas constant

j = flux, constant, just inside surface of diffused system, moles cm^{-2} sec^{-1}

θ_1 = a theta function

$t_{\frac{1}{2}}$ = half rise time, $J_t/J_\infty = \frac{1}{2}$, sec.

J_I = input current for bi-electrode experiment, current cm^{-2}

J_0 = output current, current cm^{-2}

$\Delta_1 = \int_0^{t_1} J_I dt - \int_0^{t_1} J_0 dt$, coulombs cm^{-2}

t_1 = sometimes at steady state after diffusion commences, sec.

$\Delta_2 = \int_0^{t_2} J_I dt - \int_0^{t_2} J_0 dt$, coulombs cm^{-2}

t_2 = sometime at steady state after diffusion is equalized at both sides of diffused membrane, sec.

= surface coverage, coulombs cm^{-2}

δ' = undefined thickness within diffused membrane, cm

$\Delta_3 = \int_t^\infty J_I dt - \int_{t_2}^\infty J_0 dt$, coulombs cm^{-2}, integrals extending to ∞ after outdiffusion commences at both sides of the diffused membrane, coulombs cm^{-2}

t_o = constant, time to establish initial profile in diffused system, sec.

c_e = final, equilibrium concentration in diffused system, moles cm^{-3}

K = constant in radiation treatment of diffusion

Δt = time to remove oxide film by cathodic charging, sec.

e = equilibrium electrode potential, volts

η = overpotential, volts

F = the Faraday

NOMENCLATURE

f_{H_2} = fugacity, atmospheres, of hydrogen gas
μ = chemical potential
f_H = 2-D partition function for H gas
U = potential energy
ΔH_{des} = heat of desorption
m_H = mass of hydrogen atom
m_m = mass of hydrogen molecule
I = moment of inertia of hydrogen molecule
A = surface area
N_S = total number of sites for adsorption on the electrode surface
k = Boltzmann's constant
υ = vibration frequency
$A' = 8.9 \times 10^{-40}$ for f_{H2} in atmospheres
β = symmetry factor $\sim 1/2$
k_1 = forward rate constant for first reaction
k_2 = forward rate constant for second reaction
k_{-1} = reverse rate constant
k_{-2} = reverse rate constant
$'$ = statistical temperature = $e^{-1/kT}$
= absolute activity = $e^{\mu/kT}$
$f'_H(T)$ = 3-D translational partition function for hydrogen atoms
αN = total number of sites for protons inside metal
N_H = number of protons
V = volume
ρ = nuclear spin weight
w_H = potential energy of protons inside metal relative to the state of infinite dispersion outside
w_{HH} = constant
θ'' = fraction of sites occupied by hydrogen atoms
χ_d = energy difference between the H_2 molecule in its lowest state and two free atoms in their lowest states

NOMENCLATURE

h = Planck's constant

π = spreading pressure, ergs cm^{-2}

F^{ads} = Gibbs free energy of adsorption

i = current density

i_o = exchange current density

Γ' = grand partition function

K' = constant

W = an energy and Poisson's ratio

P_{cr} = critical pressure within a crack for its propagation

c_{cr} = critical concentration

Y = Young's modulus

ΔH_s = heat of solution of hydrogen in a metal

γ = surface energy per unit area

HYDROGEN ENTRY and ACTION in METALS

1
Introduction

AREAS AFFECTED

The field of hydrogen in metals has attracted curiosity for at least 100 years. At room temperature, hydrogen diffuses through iron with a diffusion coefficient of roughly 10^{-5} $cm^2 sec^{-1}$, as fast as an ion in aqueous solution. A stretch of the imagination is required to comprehend how this can happen, and it is still not completely understood.

There was a large amount of work done (about 1,000 references) in the 1930s and 1940s on thermodynamic properties of metal hydrogen systems, solubility as a function of temperature, pressure, and so on (Smith, 1948). Smith was of the opinion that hydrogen existed in the metals in "rifts," i.e., did not dissociate and go into the bulk of the metals, existing as partially shielded protons, as is currently believed to be the case.

The field has always had its practical as well as its fundamental side. One study, though not entirely of interest because it involved hydrogen diffusion through non-metals, was the treatment (Daynes, 1920) of hydrogen diffusion through rubber membranes of material found in "air ships" of that time. From these considerations came Daynes' now well-known time lag approximation. The diffusion of hydrogen through rubber must have been a serious problem then. No less serious are the space age issues of hydrogen embrittlement, catalysis, and problems associated with the bringing about of the hydrogen economy, a term coined as early as 1971 and more noticeably in 1972 (Bockris, 1972).

Embrittlement is important in literally all applications of high strength steels because, as a general rule, the higher the strength of a steel, the greater its susceptibility to hydrogen embrittlement. About one ocean-going ship per year is lost due to embrittlement, as hydrogen evolution is an unwanted side reaction of the cathodic protection used in these vessels. The diagnostic testing of steel parts for embrittling hydrogen has recently taken a step forward. The technology of the barnacle electrode, a nondestructive testing device for hydrogen in metals (Deluccia and Berman, 1981), has been made available after a long period in development.

There are thoughts that, in the future, internal combustion engines will be powered by hydrogen stored in the form of metal hydrides. This represents a multitude of problems such as hydrogen embrittlement, storage methods, hydrogen diffusion problems, high pressure problems, and so on.

Fuel cells in combination with nuclear reactors, photovoltaics, or other methods of making hydrogen are believed to offer a solution to the electric utility problem, as organic fuels become scarce; and two facilities employing low temperature ($\sim 180°$ C phosphoric acid electrolyte) hydrogen-oxygen fuel cell systems are already under construction. Better and cheaper catalysts are needed for fuel cells; in particular, for the oxygen reduction side which is more irreversible than the hydrogen oxidation side. Considerations following from energetics data suggesting avenues of approach to better catalysts for both sides of the fuel cell will be discussed.

INTRODUCTION

   This book is intended for those just entering into the field of hydrogen in metals, and the mathematics presented in the text and appendixes will be derivations in more detail than are found elsewhere. Some background in statistical mechanics, electrochemistry, and differential equations is required.

   The material in chapters 2, 3, and 4 is preparatory for the last four chapters. For a quick overview of the book, chapters 2, 3, and 4 may be skipped over.

2
Diffusion Studies: The Bi-Electrode Technique

INTRODUCTION

The following diffusion equation, Fick's Second Law, governs what occurs in diffusion experiments. It is given in one dimension as:

$$\frac{\partial^2 c}{\partial x^2} - \frac{1}{D} \frac{\partial c}{\partial t} = 0 \qquad \qquad 2\text{-}1$$

where c is the concentration in moles cm^{-3}, x the distance in cm, D the diffusion coefficient in $cm^2 sec^{-1}$, and t the time in sec.

The theory of heat transfer follows, from the same equation, with different variables:

$$\frac{\partial^2 T}{\partial x^2} - \frac{1}{\kappa} \frac{\partial T}{\partial t} = 0 \qquad \qquad 2\text{-}2$$

where T is temperature, and κ is called the thermal diffusivity.

Good sources for the theory of diffusion are The Mathematics of Diffusion (Crank, 1956) and the Conduction of Heat in Solids (Carslaw & Jaeger, 1959). For Laplace transforms and various integrals, the Handbook of Mathematical Functions (Abramowitz & Stegun, 1965) is a good beginning.

Various solutions to the diffusion equation, and experiments, are presented in the text and appendixes. These are, for the most part, works which have been done within the last 15 to 20 years, and are of interest for the discussion in the text.

THE BI-ELECTRODE TECHNIQUE

Since the bi-electrode technique was originated in the recent past, it is still possible to include descriptions of most of the interesting experiments, in not too long a space. We will present these experiments in historical order, starting with the first bi-electrode experiments (Devanathan & Stachurski, 1962) which involved a modification of an earlier experiment (Frumkin & Aladjalova, 1944).

Devanathan and Stachurski were interested in hydrogen diffusion in iron, in an effort to study the embrittling characteristics of certain cadmium electroplating solutions. Since iron corrodes in the electrochemical environment of the bi-electrode technique, it was proposed to cover the more anodic side of the iron membrane with palladium. The first experiment, thus, was a study of hydrogen diffusion through palladium to assess the practicality of a thin layer of palladium for hydrogen diffusion. Figure 2.1 shows the original cell. In the figure, the glass joints are as labeled, and R_1 and R_2 are two identical calomel reference electrodes; G_1 and G_2 are glass floats; A_1 and A_2 are counter electrodes, and the membrane (working electrode) labeled cathode and anode, is actually ground, as with all deFord type three electrode (working, counter, and reference) circuits; L_1 and L_2 are Luggin capillaries, and A_1 is a piece of platinum wire,

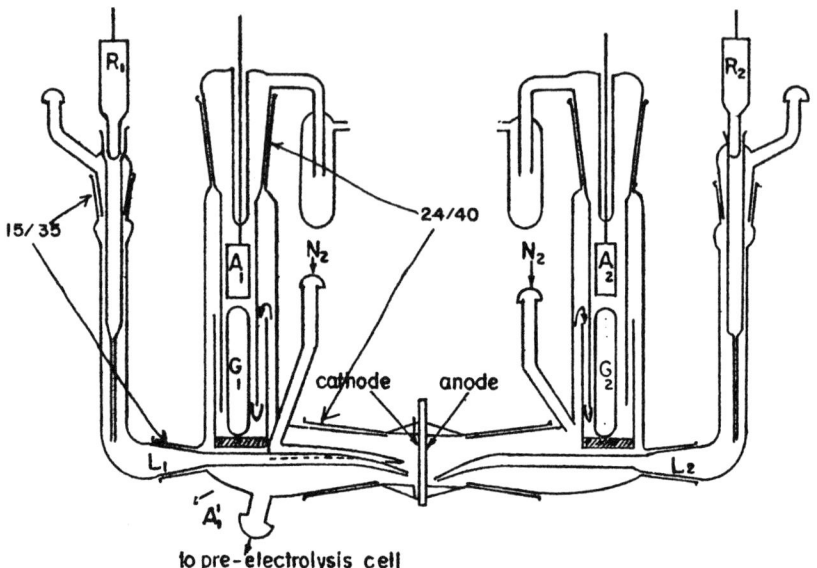

Fig. 2.1. Bi-Electrode Cell

Source: Stachurski, J.Z. and Devanathan, M.A. 1962. "The Adsorption and Diffusion of Electrolytic Hydrogen in Palladium," Proc. Roy. Soc., A270: 90. Reprinted by permission.

penetrating the cell wall for fast cathodic protection of the working electrode on the input side in later experiments with iron, as the electrolyte enters the cell. The gaskets on either side of the palladium membrane were flat pieces of teflon. The electrolyte was de-aerated and pre-electrolyzed 0.1N NaOH, and palladium membrane thicknesses were .006 to .051 cm.

At the beginning of the experiment, both sides of the membrane were potentiostated at the potential of the reversible hydrogen electrode (RHE), keeping the palladium hydrogen free. Other workers have used potentials between 0 and +300 mV to the RHE for this purpose. More positive potentials are needed for experiments in the undervoltage region.

THE BI-ELECTRODE TECHNIQUE

Charging with hydrogen at the cathodic side was accomplished by bringing the potential to -600 mV, RHE, producing a current density of about 10^{-5}A cm^{-2}. The anodic side was kept at the potential of the RHE oxidizing the hydrogen as it arrived at this surface and resulting in a current transient depicted in figure 2.2, a quantitative measure of the hydrogen permeating the metal membrane.

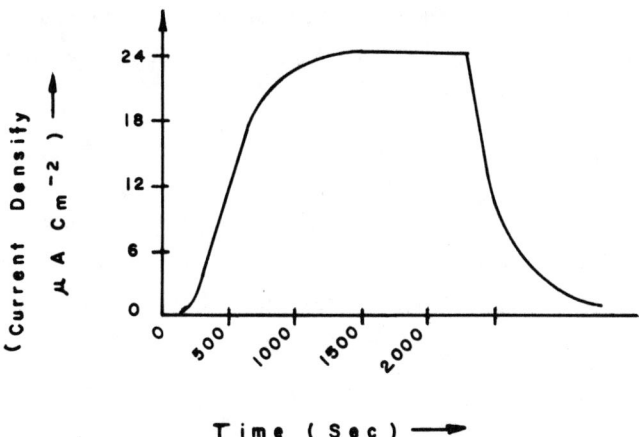

Fig. 2.2. Typical Anodic Transient for Hydrogen Diffusion in Palladium

Source: Stachurski, J.Z. and Devanathan, M.A. 1962. "The Adsorption Diffusion of Electrolytic Hydrogen in Palladium," Proc. Roy. Soc. A270: 90. Reprinted by permission.

After the trace of the current at the anodic side reached a steady value, the potential at the more negative side was returned to the RHE potential, extracting all the hydrogen again. It is worth mentioning that most later investigators used galvanostatic charging on the more negative side.

In the first experiment, a rather obscure method for the mathematical model was used. A better model, with the same characteristics, was found later (McBreen, Nanis, & Beck, 1966). The latter is a very simple model involving the equilibrium boundary condition corresponding to the concept of thermal equilibrium of heat transfer. In this, it is assumed that any variation of surface coverage with hydrogen will result in an instantaneous changes of c_o, the concentration of H just inside the metal surface. This boundary condition is expressed as:

$$c(0, t) = c_o, \text{ constant}, x = 0, t > 0 \qquad 2\text{-}3$$

Formally, the problem is: diffusion through a thin foil, equilibrium assumed at the input surface. Taking input surface as $x = 0$, and output surface at $x = \delta$, we have, as the rest of the boundary conditions:

$$c(x,o) = 0, 0 < x < \delta, t > 0 \qquad 2\text{-}4$$

$$c(\delta, t) = 0, x = \delta, t > 0 \qquad 2\text{-}5$$

and, if $\bar{c}$ denotes Laplace Transform:

$$\frac{d^2 \bar{c}}{dx^2} - \frac{p}{D} \bar{c} = 0, \qquad 2\text{-}6$$

a regular differential equation with solution:

$$\bar{c} = ae^{-qx} + be^{qx} \qquad 2\text{-}7$$

where q is $(p/D)^{\frac{1}{2}}$ and,

$$\bar{c} = c_o/p, x = 0 \qquad 2\text{-}8$$

$$\bar{c} = 0, x = \delta \qquad 2\text{-}9$$

then to evaluate a and b:

$$0 = ae^{-q\delta} + be^{q\delta} \qquad 2\text{-}10$$

$$a = -be^{2q\delta} \qquad 2\text{-}11$$

THE BI-ELECTRODE TECHNIQUE

$$\frac{c_o}{p} = -be^{2q\delta} + b \qquad 2\text{-}12$$

$$b = \frac{c_o}{p} \frac{e^{2q\delta}}{1-e^{2q\delta}} \qquad 2\text{-}13$$

$$a = -\frac{c_o}{p} \frac{e^{2q\delta}}{1-e^{2q\delta}} \qquad 2\text{-}14$$

and from Eq. 2-7:

$$\bar{c} = -(c_o/p) \frac{e^{q\delta} e^{-qx}}{e^{-q\delta}-e^{q\delta}} + (c_o/p) \frac{e^{-q\delta} e^{qx}}{e^{-q\delta}-e^{q\delta}} \qquad 2\text{-}15$$

proceeding without the expression for c (x, t), we take:

$$\bar{J} = -D\,(\partial c/\partial x)_{x=\delta} \qquad 2\text{-}16$$

$$\bar{J} = \frac{c_o D^{\frac{1}{2}}}{p^{\frac{1}{2}}} \frac{2}{e^{-q\delta} + e^{q\delta}} \qquad 2\text{-}17$$

or:

$$\bar{J} = \frac{c_o D^{\frac{1}{2}}}{p^{\frac{1}{2}}} \frac{1}{\cosh p^{\frac{1}{2}} \delta/D^{\frac{1}{2}}} \qquad 2\text{-}18$$

and from the transform:

$$\frac{1}{p^{\frac{1}{2}}} \frac{1}{\cosh p^{\frac{1}{2}}} \leftrightarrow \hat{\theta}_2 (\tfrac{1}{2},\, t) \qquad 2\text{-}19$$

where:

$$\hat{\theta}_2 (\tfrac{1}{2},\, t) = \frac{2}{(\pi t)^{\frac{1}{2}}} \sum_{m=1}^{\infty} (-1)^m e^{-\frac{1}{t}(m-\frac{1}{2})^2} \qquad 2\text{-}20$$

This corrects a mistake in Fullenwider, 1975 where θ_o was listed as a solution to this problem. We get:

$$J_t = \frac{2D^{\frac{1}{2}} c_o}{(\pi t)^{\frac{1}{2}}} \sum_{m=1}^{\infty} (-1)^m e^{-(m-\frac{1}{2})^2 \delta^2/Dt} \qquad 2\text{-}21$$

and taking:

$$J_\infty = \frac{D\,c_o}{\delta} \qquad 2\text{-}22$$

we have:

$$\frac{J_t}{J_\infty} = \frac{2\delta}{(\pi Dt)^{1/2}} \sum_{m=0}^{\infty} (-1)^m e^{-(2m+1)^2 \delta^2 / 4Dt} \qquad 2\text{-}23$$

where J_t/J_∞ is the fractional attainment of steady state of the flux through the thin metal foil. In the past (McBreen, Nanis, & Beck, 1966), Eq. 2-23 has been used in the approximation:

$$\frac{J_t}{J_\infty} = \frac{2\delta}{(\pi Dt)^{1/2}} e^{-\delta^2 / 4Dt} \qquad 2\text{-}24$$

By an alogarithm, Eq. 2-24 can be solved for D:

$$D = 0.14 \frac{\delta^2}{t_{1/2}} \qquad 2\text{-}25$$

where $t_{1/2}$ is the half rise time, and since:

$$J_\infty = \frac{D c_o}{\delta} \qquad 2\text{-}26$$

the steady state concentration profile is assumed to be linear, and we have for c_o:

$$c_o = \frac{\delta J_\infty}{D} \qquad 2\text{-}27$$

Thus, we have access to all the variables through experiment, though it is still possible to obtain more information. The diffusion coefficient of hydrogen in palladium was determined to be 1.27×10^{-7} cm^2 sec^{-1}, at room temperature.

The second experiment with the bi-electrode technique (Devanathan, Stachurski, & Beck, 1963) was, as planned, with iron membranes, 0.77mm thick.

The cathodic side of the membrane was protected by the cathodic current; and, on the anodic side, a thin film (1,000 Å) of shiny palladium was plated on prior to the experiments. The preparation of the palladium plating solution finds mention only in obscure sources and will be described here.

THE BI-ELECTRODE TECHNIQUE

One gram of palladium chloride ($PdCl_2$) was added to 100 ml of purified water at 80° C. Sodium nitrite ($NaNO_2$) crystals were added until the palladium chloride completely reacted to form a yellow solution of $Na_2[Pd(NO_2)_4]$. One ml of this solution was added to one liter of 0.2 N sodium hydroxide solution. To plate, current density of 4×10^{-4} A cm^{-2} was used, for one hour, with agitation, resulting in a bright palladium film 1,000 Å thick.

Below are described several applications of Devanathan and Stachurski's permeation technique. The first work (Devanathan & Stachurski, 1962) described the embrittling properties of various cadmium electrolytic plating solutions investigated with the bi-electrode technique, the amount of hydrogen diffusing through the metal membranes being used as an indicator at a cathodic current density of 8.1 mA cm^{-2}. The electrolyte was 0.2 N sodium hydroxide on the cathodic side. The anodic electrolyte was the palladium plating solution, de-aerated with purified nitrogen (agitation) as the palladium coating was put on.

The cadmium electroplating solutions decreased as follows in order of amount of diffusing hydrogen:

standard cyanide bath
> fluoborate bath
> amino butyrate bath

Information of this type is useful for such organizations as the Naval Air Engineering Center, who plate steel parts such as airplane landing gears with cadmium for protection from ocean environments and, more recently, bake plated parts at 300° F to drive out the embrittling hydrogen.

The diffusion coefficients for pure iron and 4340 steel (high strength) were determined as, respectively, 8.3×10^{-5} cm^2 sec^{-1}, and 2.0×10^{-7} cm^2 sec^{-1} at room temperatures.

A point of interest is the preparation of the iron membranes. These were annealed prior to experiment, in hydrogen, for two hours at 600° C. It was stated that "pre-saturation with hydrogen is indispensable for the attainment of uniform permeation" (Devanathan & Stachurski, 1962).

The 4340 steel membranes were presaturated electrolytically in 0.1N H_2SO_4 at a current density of 10 mA cm^{-2} for six hours. Annealing was avoided here because 4340 steel is tempered at about 500° F.

These pretreatment procedures will become important later when charging mode (potentiostatic versus galvanostatic) is discussed.

The next work (Beck, Bockris, McBreen, & Nanis, 1966) was an interesting study with hydrogen diffusion in iron, zone refined iron, single crystal armco iron, and 4340 steel. The iron membranes were annealed as by Devanathan and Stachurski, but the steel specimens were used as received. Hydrogen diffusion was studied as a function of stress, temperature, and dissolved hydrogen concentration. Here, already, is reference to anomalies in permeation behavior, avoided by working at cathodic current densities less than 15 mA cm^{-2}.

The diffusion coefficient for hydrogen in iron was found to be 6.25 x $10^{-5} cm^2 sec^{-1}$ (comparing well with the previously cited value); for zone refined iron 6.05 x $10^{-5} cm^2 sec^{-1}$; and single crystal iron (100 direction) showed an increase of 30 percent to 8.25 x $10^{-5} cm^2 sec^{-1}$, at room temperature. The increase between the first two diffusion coefficients, which are essentially the same, and that for the 100 direction in single crystal iron was explained as being due to straight path diffusion in the single crystal case as opposed to a tortuous path in the polycrystalline situation. These findings were taken as proof that diffusion occurred through bulk metal and not along cracks or grain boundaries.

THE BI-ELECTRODE TECHNIQUE

The diffusion coefficient for 4340 steel, including temperature dependence, was found to be:

$$D = 4.47 \times 10^{-2} e^{-9220/RT} \text{ cm}^2 \text{sec}^{-1} \qquad 2\text{-}28$$

Of more interest was the determination of permeation as a function of stress. Experiments with Armco iron and 4340 steel showed an increase in permeation with applied tensile stress. In addition, the diffusion coefficients were unaltered. Thus, solubility alone was the parameter of change.

The investigators found a relation (Beck et al., 1966):

$$\frac{\partial \ln (c_{H,\sigma H}/c_{H,o})}{\partial \sigma_H} = \frac{\bar{V}}{RT} \qquad 2\text{-}29$$

or:

$$c_{H,\sigma_H} = c_{H,o} e^{\bar{V}_H, \sigma_H /RT} \qquad 2\text{-}30$$

for the partial molar volume, $\bar{V}$, where $c_{H,\sigma H}$ is the concentration at stress σ_H, $c_{H,o}$ the concentration at zero stress, and σ_H the hydrostatic stress. The plots of J_σ/J_o all lay on the same curve. From this, it was possible to calculate:

$$\bar{V} = 0.39 \text{ cm}^3/\text{gm H} \qquad 2\text{-}31$$

From experiments involving compressive as well as tensile stress, it was found (Bockris, Beck, Genshaw, Subramanyan, & Williams, 1971):

$$\bar{V} = 2 \text{ to } 3 \text{ cm}^3/\text{gm H} \qquad 2\text{-}32$$

for Armco iron and 4340 steel, corrected (Beck et al., 1966) partly by a factor of 2.303, the natural log conversion factor (an error on the part of Beck et al.), and also by an increase in accuracy.

In addition, it was calculated (Bockris et al., 1971) that the total energy change in introducing one mole of H into iron is 59 kcal. This had been thought by some to be too large (Flanagan & Oates, 1972). This may not be an excessive value, however, for the following reason: The heat of sublimation of iron which is equivalent to the cohesive energy is 96.7 kcal mole^{-1} at 25° C. Upon introduction of one mole of H into iron lattice, the cohesive energy of the given amount of iron (it could be very large or very small like one g atom of Fe) would be reduced by 59 kcal.

The strain produced by the entry of an H atom into the iron lattice was calculated on the basis of a sample volume with all the interstices filled with atomic H. This was simply a model and not meant to be achieved in practice. Such a system would blow apart because the structure of 1 gm atom of Fe could accommodate 3 g atom of H. The cohesive energy of Fe 96.7 kcal (gm atom)$^{-1}$ would be exceeded by the strain energy of 3 g atom of H = 3 x 59 kcal.

A BI-ELECTRODE EXPERIMENT WITH THE HYDROGEN PALLADIUM SYSTEM

We will now describe a bi-electrode experiment with the hydrogen-palladium system in the undervoltage region in which all of the diffusing hydrogen was accounted for (Fullenwider, 1975). The gasket shown in figure 2.3 was designed for this purpose. In addition to being easy to use, the gasket presses in firmly in a line seal about the diffusing region of the membrane, making it difficult to lose hydrogen through edge diffusion. All diffusing hydrogen goes into palladium as long as potentials in the undervoltage region are used.

In most galvanostatic bi-electrode experiments conducted in the past, notably with iron and iron alloys, the overpoten-

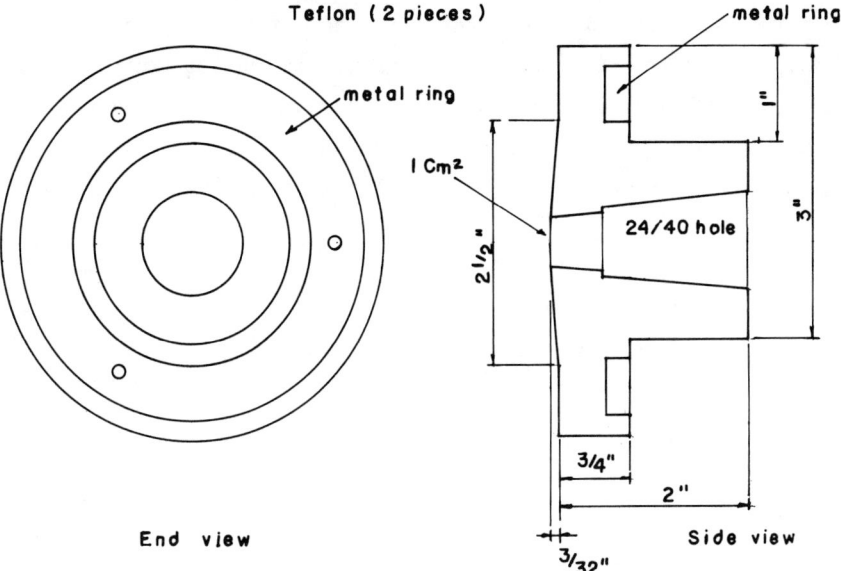

Fig. 2.3. Center Gasket of Bi-Electrode Cell

tial has ranged to high cathodic values, and the cathodic current was large. This made it impossible to distinguish between evolved hydrogen and the fraction going into the metal at the input side. What happens is that the two quantities differ by orders of magnitude. With galvanostatic experiments, the anodic current provides this information, however; but there is still loss of information, making the surface coverage determination relative, only, as will be shown.

In the undervoltage region, as previously stated, with palladium, all of the hydrogen generated at the input side goes into the membrane, resulting in ideal experimental conditions, i.e., here there are µA's going in, and µA's coming out, making it possible to add and subtract the various integrals of current. Data was collected with potentiostatic charging. Another quantity which can be obtained here is the absolute coverage. Until this experiment, relative coverage only had been measured.

First, we need a constant flux model for diffusion through a thin foil. It will be of use in the analysis of the experiment. We take, assuming that any change in surface coverage results in an instantaneous change in the flux, j, just inside the metal surface (Fullenwider, 1975):

$$-D \left(\frac{\partial c}{\partial x} \right)_{x=0} = j, \text{ constant, } t > 0 \qquad 2\text{-}33$$

$$c(\delta, t) = 0, \ x = \delta, \ t > 0 \qquad 2\text{-}34$$

$$c(x, t) = 0, \ 0 < x < \delta, \ t \leq 0 \qquad 2\text{-}35$$

Transforming:

$$-D \left(\frac{\partial \bar{c}}{\partial x} \right)_{x=0} = \frac{j}{p}, \ x = 0 \qquad 2\text{-}36$$

$$\bar{c} = 0, \ x = \delta \qquad 2\text{-}37$$

$$\bar{c} = 0, \ 0 < x < \delta \qquad 2\text{-}38$$

$$\bar{c} = a e^{-qx} + b e^{qx} \qquad 2\text{-}39$$

$$\frac{j}{p} = -D a q + D b q \qquad 2\text{-}40$$

$$0 = a e^{-q\delta} + b e^{q\delta} \qquad 2\text{-}41$$

$$a = -b e^{2q\delta} \qquad 2\text{-}42$$

$$\frac{j}{p} = -b e^{2q\delta} + bq \qquad 2\text{-}43$$

$$b = \frac{-j}{pq} \frac{1}{e^{2q\delta} + 1} \qquad 2\text{-}44$$

$$a = \frac{j}{pq} \frac{e^{2q\delta}}{e^{2q\delta} + 1} \qquad 2\text{-}45$$

$$\bar{c} = \frac{j \, e^{2q\delta} \, e^{-qx}}{pq \, (e^{2q\delta} + 1)} - \frac{j \, e^{qx}}{pq \, (e^{2q\delta} + 1)} \qquad 2\text{-}46$$

with:

$$\bar{J} = -D \, (\partial c / \partial x)_{x = \delta} \qquad 2\text{-}47$$

$$\bar{J} = \frac{2 \, j \, e^{q\delta}}{p \, (e^{2q\delta} + 1)} \qquad 2\text{-}48$$

THE BI-ELECTRODE TECHNIQUE

$$\bar{J} = \frac{j}{p \cosh q \, \delta} \qquad \text{2-49}$$

and with the transform:

$$\frac{1}{p \cosh p^{\frac{1}{2}}} \leftrightarrow 1 - \int_0^1 \theta_1 \left(\frac{u}{2}, t\right) du \qquad \text{2-50}$$

$$\frac{J_t}{J_\infty} = 1 - \frac{4}{\pi} \sum_{m=0}^{\infty} [(-1)^m/(2m+1)] \exp[-\pi^2 (2m+1)^2 Dt/4\delta^2] \qquad \text{2-51}$$

or:

$$D = 0.38 \times \delta^2/t_{\frac{1}{2}} \qquad \text{2-52}$$

and:

$$c_o = \frac{\delta \, J_\infty}{D} \qquad \text{2-53}$$

It can be seen that the estimate of the diffusion coefficient will have greater numerical values with this model. It should be noted that all determinations of the diffusion coefficient with the bi-electrode technique involve surface effects. An important check on the diffusion coefficient would involve the nuclear magnetic resonance or slow neutron scattering method, where surface properties play no part. To date, however, there is no working model for these experiments.

Now, a few words about the experimental technique. The cell was of the type pictured in figure 2.1, a bi-electrode cell, and the gasket was as in figure 2.3. This gasket pressed in firmly about the diffusing portion of the palladium membranes preventing "edge" diffusion. The membranes were 7 x 10^{-3} cm in thickness. The palladium pieces, dimensionally 1 x 3 inches, were annealed at 900° C in an argon atmosphere for two hours prior to experiment and stored under argon until used. The electrolyte on both sides of the cell was 0.1 N H_2SO_4.

The potential on both sides of the cell controlled with de Ford type potentiostats. This is not the usual approach. The input side in most bi-electrode experiments is controlled galvanostatically. Potentiostatic control was used here to ensure that the electrochemical potential and associated equivalent pressure were held constant throughout any experiment.

The rationale of the method can be understood by referring to figures 2.4 to 2.6. In figures 2.4 and 2.5, typical input and output traces are shown. Prior to t = 0, both sides of the membrane are at 300 mV anodic to the reversible hydrogen potential, and no hydrogen is present; at t = 0, the input side of the membrane is brought to, in this case, a potential of 136 mV, introducing hydrogen at this side; at t_1, some time after equilibrium has been established, the output side is raised to the same potential as the input side, producing a uniform concentration profile, and at t_2, both sides are again taken to 300 mV, extracting all of the hydrogen.

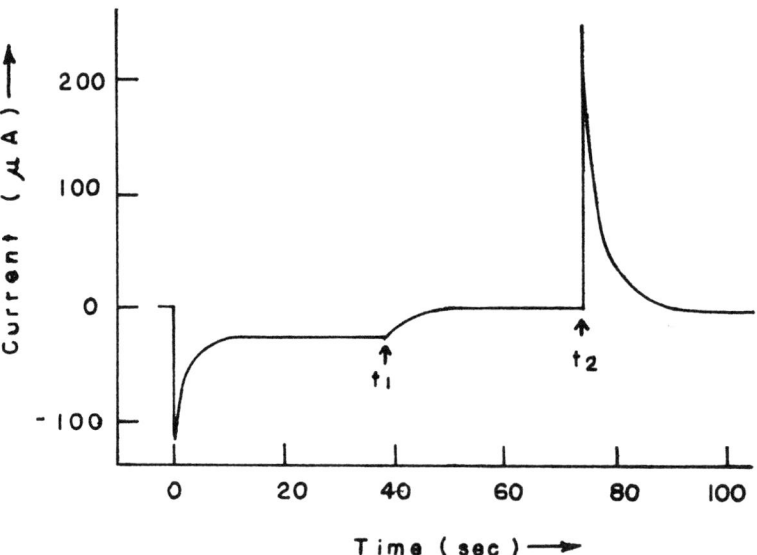

Fig. 2.4. Input Trace (current balance experiments)

Reprinted with permission of the publisher, The Electrochemical Society, Inc.

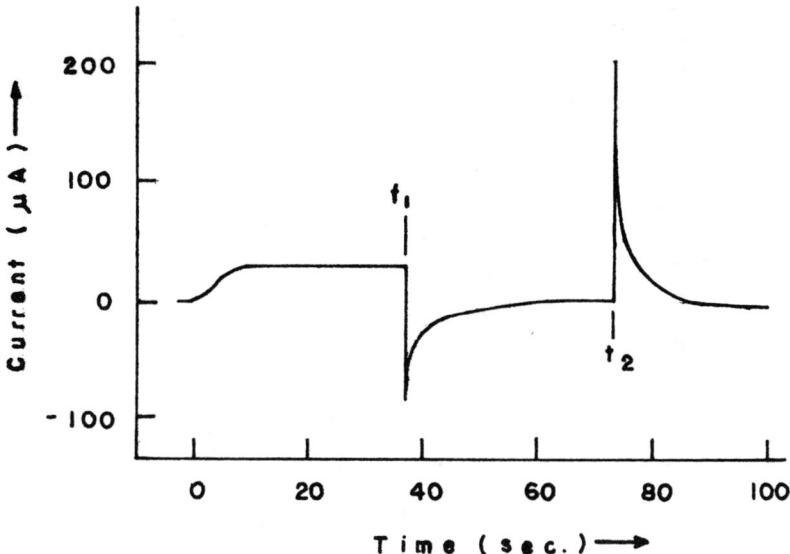

Fig. 2.5. Output Trace Corresponding to Figure 2.4.

Now assuming, for a moment, a steady-state profile at t_1 of the type shown in figure 2.6 in the event that material balance is attained, it is possible to define some quantities. Again referring to figures 2.4, 2.5, and 2.6, we have:

$$\int_0^{t_1} J_I dt - \int_0^{t_1} J_O dt = \Delta_1 = \Gamma + c_o \delta' + c_o \delta/2 \qquad 2\text{-}54$$

and

$$\int_0^{t_2} J_I dt - \int_0^{t_2} J_O dt = \Delta_2 = 2\Gamma + c_o (\delta' + \delta) \qquad 2\text{-}55$$

or

$$\Delta_2 - 2\Delta_1 = -c_o \delta' \qquad 2\text{-}56$$

where δ' is some undefined distance into the membrane, δ is the thickness of the membrane, J_I is the input current, J_O is the output current, Γ is the surface coverage, and c_o is the maximum hydrogen concentration within the membrane. Also, we have a quantity:

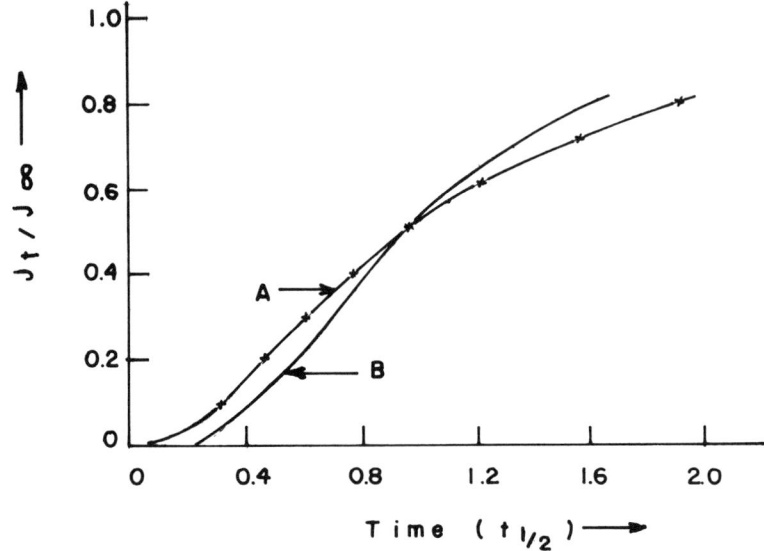

Fig. 2.6. Equilibrium A and Constant Flux Profile B Normalized about the Half Rise Time. Data for the Hydrogen - Palladium System.

Reprinted by permission of the publisher, The Electrochemical Society, Inc.

$$\Delta_3 = \int_{t_2}^{\infty} J_I dt + \int_{t_2}^{\infty} J_0 dt \qquad 2\text{-}57$$

which should be equal to Δ_2, the total amount of hydrogen in the membrane at saturation.

If the quantity $\Delta_2 - 2\Delta_1$ is negative, the profile is of the type shown in figure 2.7 (D increases with increasing concentration); if it is zero, the profile is linear (D is independent of concentration); and if it is positive, the profile is concave upward (D decreases with increasing concentration).

In the case that the diffusion coefficient is independent of concentration, the following relation holds for $t < t_1$ (Schmidt & Siegenthaler, 1970):

THE BI-ELECTRODE TECHNIQUE

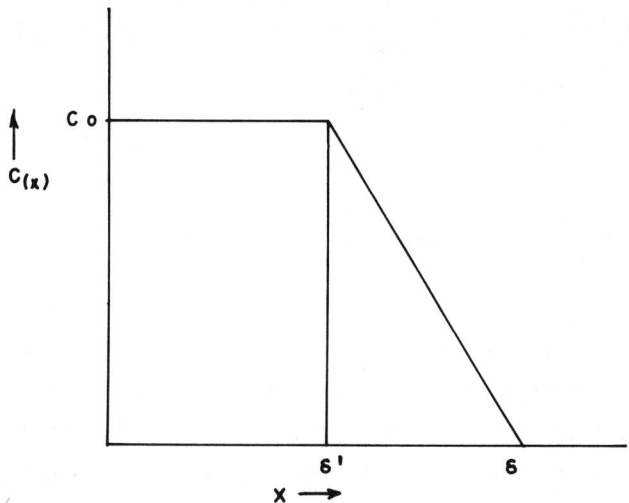

Fig. 2.7. Hypothetical Concentration Profile Indicating Functionality of Dependence of Diffusion Coefficient on Concentration.

Reprinted by permission of the publisher, The Electrochemical Society, Inc.

$$\int_0^t J_I dt - \int_0^t J_0 dt = \Gamma + \delta^2 J_0/2!D + (\delta^4/4!D^2) dJ_0/dt$$
$$+ (\delta^6/6!D^3) d^2 J_0/dt^2 + \ldots \qquad 2\text{-}58$$

or at t_1

$$\int_0^{t_1} J_I dt - \int_0^{t_1} J_0 dt = \Gamma + \delta^2 J_0/2D \qquad 2\text{-}59$$

where D is to be determined from the shape of the output transient by suitable mathematical analysis.

The two expressions in Eq.'s 2-23 and 2-51 can best be compared according to the value of Dt/δ^2 corresponding to $J_t/J_\infty = 0.5$. We will call this quantity $Dt_{\frac{1}{2}}/\delta^2$, and we have from Eq. 2-25:

$$Dt_{\frac{1}{2}}/\delta^2 = 0.14 \qquad \qquad 2\text{-}60$$

and from Eq. II-52:

$$Dt_{\frac{1}{2}}/\delta^2 = 0.38 \qquad \qquad 2\text{-}61$$

respectively.

RESULTS AND DISCUSSION

Figure 2.8 shows plots of Eqs. 2-23 and 2-51, normalized about $t_{\frac{1}{2}}$, together with a typical experimental output transient. The experimental data can be seen to fit Eq. 2-51. Previous investigators (Devanathan & Stachurski, 1962) used a concentration step function model in the analysis of their data. This would result in an underestimation of the value of the diffusion coefficient.

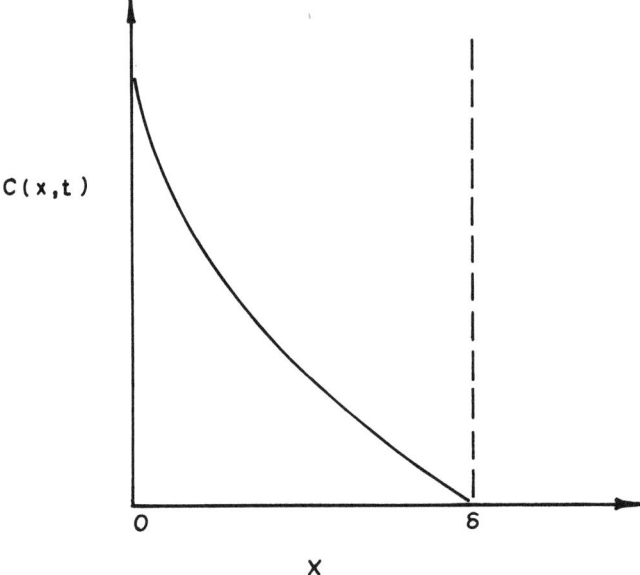

Fig. 2.8. Concentration Profile in Concentration Dependent Region

THE BI-ELECTRODE TECHNIQUE 23

In table 2.1 is listed a typical set of values of the quantities Δ_1, Δ_2, and Δ_3 at various potentials. The difference $\Delta_2 - 2\Delta_1$ (a sensitive indication of the shape of the concentration profile through the thin membrane) is positive for all but the most anodic of potentials where it is zero, i.e., the diffusion coefficient decreases with increasing concentration. Also, Δ_2 agrees fairly well with Δ_3. The decrease of the diffusion coefficient can easily be rationalized as being due to the blocking of diffusional pathways at higher concentrations. It is interesting that the previous results (Namboodhiri & Nanis, 1973) showed an increase of the diffusion coefficient of hydrogen in iron with concentration.

Also listed in table 2.1 are absolute values of Γ calculated from the relation, following from Eqs. 2-54 and 2-59:

$$\Gamma = \Delta_1 - \delta^2 J_0 / 2D \qquad 2\text{-}62$$

where the difference $\Delta_2 - 2\Delta_1$ is zero and the relations hold. It is not possible to obtain Γ by simple analysis at more negative potentials. Γ can be seen to be approaching a monolayer at about 210 mV.

The diffusion coefficient in the linear region was calculated from Eq. 2-61 to be 4.5×10^{-7} cm^2 sec^{-1}, slightly larger than values previously reported which were in the range $(1.3-3.1) \times 10^{-7}$ cm^2 sec^{-1}. We can only conclude that all previous determinations of the diffusion coefficient were either made in the concentration dependent region or involved an incorrect mathematical model.

This experiment worked with the hydrogen-palladium system. Such experiments with the hydrogen-iron system would be of more interest. Since iron corrodes in the undervoltage region, however, this seems, at present, impossible.

Recently (Early, 1978), it has been reported that the constant flux model applies to galvanostatic charging of

Table 2.1. Various Integrals, $\Delta_2 - 2\Delta_1$, and Absolute Coverages

E, mV, RHE at input surface	Δ_1, coulombs cm^{-2}	Δ_2, coulombs cm^{-2}	Δ_3, coulombs cm^{-2}	$\Delta_2 - 2\Delta_1$, coulombs cm^{-2}	Γ, coulombs cm^{-2}
74	1.8×10^{-2}	6.8×10^{-2}	7.4×10^{-2}	3.2×10^{-2}	
93	7.5×10^{-3}	2.9×10^{-2}	3.5×10^{-2}	1.4×10^{-2}	
117	4.0×10^{-3}	1.1×10^{-2}	1.3×10^{-2}	3.0×10^{-3}	
133	2.4×10^{-3}	7.6×10^{-3}	7.7×10^{-3}	2.8×10^{-3}	
154	1.6×10^{-3}	3.5×10^{-3}	3.4×10^{-3}	3.0×10^{-4}	
174	8.9×10^{-4}	1.9×10^{-3}	1.7×10^{-3}	1.0×10^{-4}	
197	4.7×10^{-4}	9.7×10^{-4}	8.0×10^{-4}	3.0×10^{-5}	
212	2.1×10^{-4}	4.2×10^{-4}	5.0×10^{-4}	0	1.3×10^{-4}
237	8.5×10^{-5}	1.9×10^{-4}	1.8×10^{-4}	0	6.0×10^{-5}

palladium with hydrogen. This was the first time that information was available for both potentiostatic (Fullenwider, 1975) and galvanostatic (Early, 1978) charging of the same metal - palladium. This brings us to the topic of mode of charging.

Here we have a problem area which has never been discussed and is often covered up by electrode pretreatment. Electrodes, or membranes, for example, are often annealed in hydrogen, plated with palladium, pulsed, etched, etc.

What happens with galvanostatic charging is that there results a slow transient followed by a series of fast reproducible transients; and with potentiostatic charging, one gets only a series of fast identical transients. Examples exist for potentiostatic charging of palladium (Devanathan & Stachurski, 1962), galvanostatic charging of iron (Wach, 1971), galvanostatic charging of platinum (Gileadi, Fullenwider, & Bockris, 1966), and galvanostatic data for palladium (Early, 1978). With galvanostatic charging, the electrode potential on the input side usually rises slowly to high cathodic values during the course of the experiments, thus making the results questionable. This behavior is the central issue in controversy, some workers (Wach, 1971) being of the opinion that surface effects causes the overpotential shift, and others (Pressouyre & Bernstein, 1978) maintaining that "trapping" in the bulk of the metal causes the slow shift.

It is possible, with galvanostatic charging, depending on the metal and current density, that the potential jumps to a steady value quickly, resulting in the same behavior as the potentiostatic case; however, this behavior is relegated to more cathodic potentials and high current densities.

3
Diffusion Studies: Barnacle Electrode

The barnacle electrode is an electrochemical device for the determination of the concentration of hydrogen in metal pieces. It is introduced here because it illustrates some of the basic diffusion procedures and problems in the study of hydrogen in metals.

The barnacle electrode was conceived in its latest form at the Naval Air Development Center (Berman, Beck, & DeLuccia, 1974; Deluccia & Berman, 1981). In this form (Fig. 3.1), the electrode is a portable and nondestructive instrument for the in-situ diagnostic measurement of mobile, embrittling hydrogen. It can detect hydrogen in concentrations below 0.1 p.p.m. The electrode is meant to attach to a metal surface by means of a clamp and was originally of interest to the Naval Bureau of Weapons, thus the name Barnacle Electrode. The driving force for the electrode is a nickel/nickel oxide sheet electrode. When contact is made between the nickel/nickel oxide piece and the metal specimen, both in electrolyte (0.2 N NaOH), the metal specimen is brought to the potential of the nickel/nickel electrode, and a potential is set up, at the potential of the reversible hydrogen electrode, which immediately oxidizes all the hydrogen from the surface of the

BARNACLE ELECTRODE

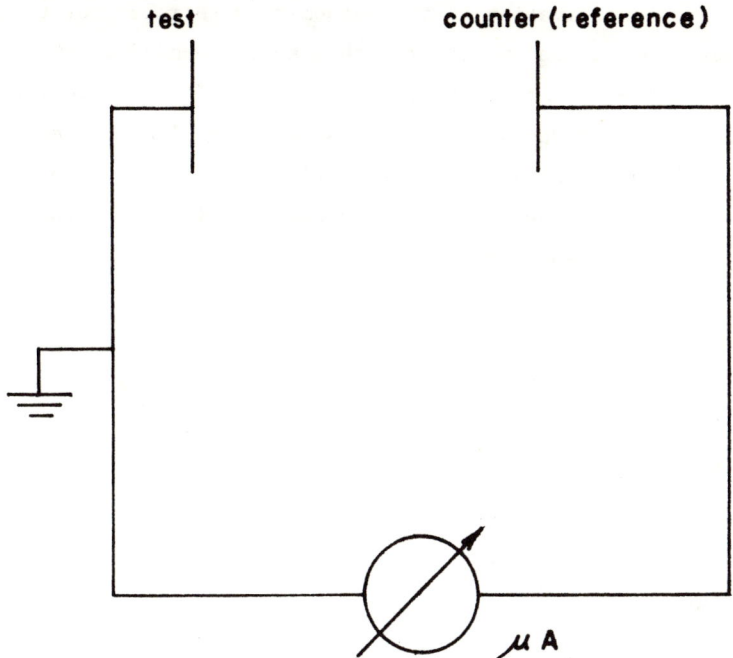

Fig. 3.1. The Barnacle Electrode

specimen and, with this, causes any hydrogen near this surface to diffuse out, resulting in a decay transient, taken down by a recorder as current. Provided the diffusion coefficient for hydrogen in the specimen is known, the recorder trace may be used directly to calculate the concentration of hydrogen in the specimen. How the diffusion coefficient can be obtained with the use of the bi-electrode technique has already been discussed.

First, we give the analysis most commonly associated with the barnacle electrode. It is assumed that the specimen has been exposed to the diffused hydrogen for a long time, producing a uniform gradient.

In the derivation of the mathematics formulas for the barnacle electrode, we start with a solid bounded by two parallel planes, and then see which of the final approximations (finite or semi-infinite) is more useful at the end.

Taking the solid to be symmetrical about $x = 0$, $-\delta < x < \delta$, we arrive at the following boundary and initial conditions (Bockris, Genshaw, & Fullenwider, 1970):

$$\frac{\partial c}{\partial x} = 0, \quad t > 0, \quad x = 0 \qquad 3\text{-}1$$

$$c = 0, \quad x = \pm \delta, \quad t > 0 \qquad 3\text{-}2$$

$$c = c_o, \quad -\delta < x < +\delta, \quad t \leq 0 \qquad 3\text{-}3$$

At this point, we can neglect the solid extending to $-\delta$, and Eqs. 3-2 and 3-3 become:

$$c = 0, \quad x = \delta, \quad t > 0 \qquad 3\text{-}4$$

$$c = c_o, \quad 0 < x < \delta, \quad t \leq 0 \qquad 3\text{-}5$$

3-1, however, still holds. This extra step shows clearly why Eq. 3-7 and Eq. 3-1 obtain. This boundary condition is known as the "impenetrable surface" condition.

Taking the Laplace transform, we get:

$$\frac{d^2 \bar{c}}{dx^2} - \frac{p \bar{c}}{D} = \frac{-c_o}{p} \qquad 3\text{-}6$$

$$\frac{d\bar{c}}{dx} = 0, \quad x = 0 \qquad 3\text{-}7$$

$$\bar{c} = 0, \quad x = \delta \qquad 3\text{-}8$$

$$\bar{c} \text{ (general)} = A e^{-qx} + B e^{qx} + \frac{c_o}{p} \qquad 3\text{-}9$$

Here we have a simple initial condition which enters into the general solution. As before:

$$q - (p/D)^{\frac{1}{2}} \qquad 3\text{-}10$$

BARNACLE ELECTRODE

From 3-7 and 3-9:

$$o = -aq + bq \qquad 3\text{-}11$$

$$a = b \qquad 3\text{-}12$$

and with 3-8:

$$(\bar{c})_{x=\delta} = o - ae^{-q\delta} + be^{q\delta} + \frac{c_o}{Dq^2} \qquad 3\text{-}13$$

Therefore:

$$a = b = \frac{-c_o}{Dq^2 e^{-q\delta}(1 + e^{2q\delta})} \qquad 3\text{-}14$$

and

$$\bar{c} = \frac{c_o}{p} - \frac{c_o e^{-qx}(1 + e^{2qx})}{pe^{-q\delta}(1 + e^{2q\delta})} \qquad 3\text{-}15$$

or:

$$\bar{c} = \frac{c_o}{p} - \frac{c_o e^{-qx}(1 + e^{2qx})}{p} \cdot \frac{e^{-q\delta}}{1 + e^{-2q\delta}} \qquad 3\text{-}16$$

It would be correct at this point to proceed directly to the expression for the transformed flux by taking $(\partial c/\partial x)_{x=\delta}$, but here we will take 3-16 and solve for $c(x, t)$ first. Using the geometric series relation:

$$\frac{a'}{1-r} = a' + a'r + a'r^2 + \ldots, \quad r^2 < 1 \qquad 3\text{-}17$$

and taking:

$$a' = e^{-q\delta} \qquad 3\text{-}18$$

and:

$$r = -e^{-2q\delta} \qquad 3\text{-}19$$

we get:

$$\bar{c} = \frac{c_o}{p} - c_o e^{-qx} \frac{(1 + e^{2qx})}{p} \sum_{m=o}^{\infty} (-1)^m e^{-(2m+1)q\delta} \qquad 3\text{-}20$$

or:

$$c = \frac{c_o}{p} - \frac{c_o}{p}\sum_{m=0}^{\infty}(-1)^m e^{-[(2m+1)\delta-x]q}$$

$$-\frac{c_o}{p}\sum_{m=0}^{\infty}(-1)^m e^{-[(2m+1)\delta+x]q} \qquad 3\text{-}21$$

and using the transforms:

$$\frac{1}{p} \leftrightarrow 1 \qquad 3\text{-}22$$

$$\frac{1}{p}e^{-kp^{\frac{1}{2}}}, \quad k \geq 0, \leftrightarrow \text{erfc}\frac{k}{2t^{\frac{1}{2}}} \qquad 3\text{-}23$$

we have:

$$c(x,t) = c_o\sum_{m=0}^{\infty}(-1)^m \text{erfc}\frac{(2m+1)\delta-x}{(4Dt)^{\frac{1}{2}}}$$

$$+ c_o\sum_{m=0}^{\infty}(-1)^m \text{erfc}\frac{(2m+1)\delta+x}{(4Dt)^{\frac{1}{2}}} \qquad 3\text{-}24$$

or:

$$c(x,t) = -c_o + c_o\sum_{m=0}^{\infty}(-1)^m \text{erf}\frac{(2m+1)\delta-x}{(4Dt)^{\frac{1}{2}}}$$

$$+ c_o\sum_{m=0}^{\infty}(-1)^m \text{erf}\frac{(2m+1)\delta+x}{(4Dt)^{\frac{1}{2}}} \qquad 3\text{-}25$$

Going on to get:

$$J = -D\,(\partial c/\partial x)_{x=\delta} \qquad 3\text{-}26$$

we use:

$$\frac{d}{du}\text{erf}(u) = \frac{2}{\pi^{\frac{1}{2}}}e^{-u^2} \qquad 3\text{-}27$$

and then:

$$\frac{J_t}{zF} = c_o\left(\frac{D}{\pi t}\right)^{\frac{1}{2}}\left[1 - 2\sum_{m=0}^{\infty}(-1)^m e^{-(m+1)^2\delta^2/4Dt}\right] \qquad 3\text{-}28$$

as an approximation

$$\frac{J_t}{zF} = c_o\left(\frac{D}{\pi t}\right)^{\frac{1}{2}} \qquad 3\text{-}29$$

BARNACLE ELECTRODE

a relation which holds for:

$$t_{max.} \geq \delta^2/4D \qquad 3\text{-}30$$

Therefore, for an armco iron membrane 0.11 mm thick with $D = 1.7 \times 10^{-5}$ cm^{2}sec^{-1}:

$$t_{max.} = 1.8 \text{ sec.} \qquad 3\text{-}31$$

and for 4340 steel, $\delta = 0.5$ mm, $D = 4.7 \times 10^{-7}$ cm^2 sec^{-1}:

$$t_{max.} = 1330 \text{ sec.} \qquad 3\text{-}32$$

a better approximation.

This is the analysis which is used in most barnacle electrode experiments. If D is known, say from a bi-electrode experiment, then it is possible to calculate c_o. If D is not known, however, all that can be deduced is the quantity $D^{\frac{1}{2}}c_o$. See figure 3.2 for an illustration of this problem.

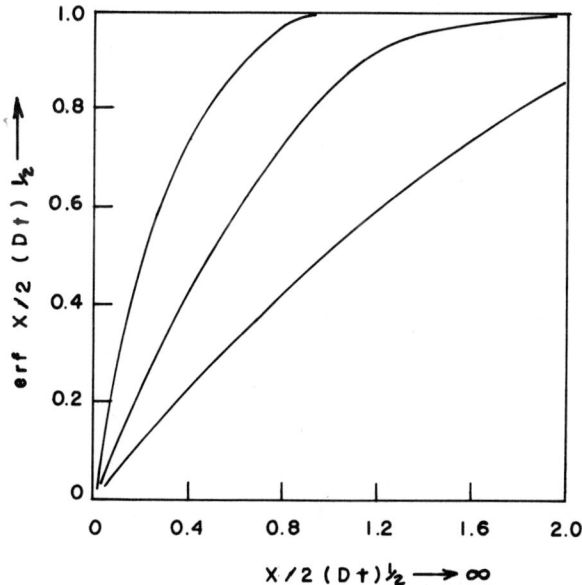

Fig. 3.2. Rediffusion for the Uniform Initial Condition

In the situation where the initial profile is not uniform, e.g., where, perhaps, the specimen has not been charged with hydrogen for a long time, or, if the Barnacle technique is used in pulsing-type experiments where the hydrogen is put in and then drawn out of the same surface region of a specimen, then the error function complement profile is useful. First, the initial condition for in-diffusion is (see Appendix A for the derivation):

$$\frac{c}{c_o} = \text{erfc} \frac{x}{(4Dt)^{\frac{1}{2}}} \qquad 3\text{-}33$$

This is the error function profile. The erfc is a well-studied function. See, for example, Fullenwider, 1974 and figure 3.3. Going on to the problem of out-diffusion from 3-33, we have;

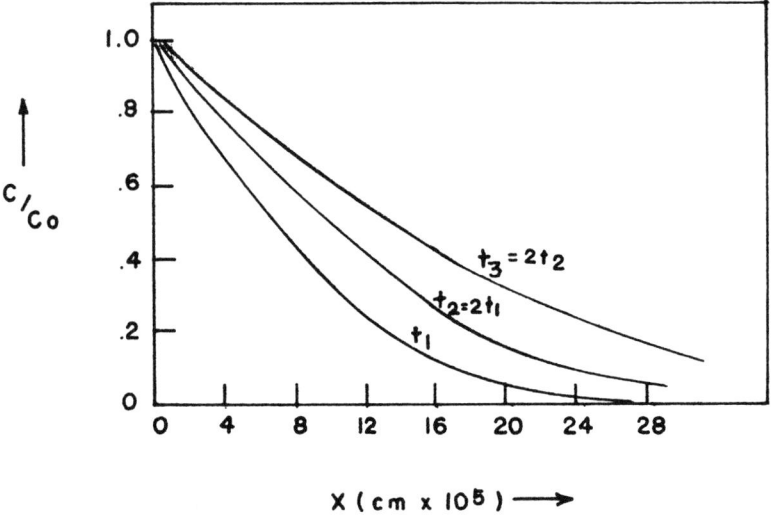

Fig. 3.3. The Error Function Complement Profile

Reprinted by permission of the publisher, The Electrochemical Society, Inc.

BARNACLE ELECTRODE

Here we consider the case of rediffusion from the error function complement distribution (Bockris et al., 1970). We will have as the initial condition, the solution derived in Appendix A.

$$c(x, t_o) = c_o \, \text{erfc} \, \frac{x}{(4Dt_o)^{1/2}} \qquad 3\text{-}34$$

Taking Eq. 3-34 as an initial condition (see Appendix B), we obtain for the problem of rediffusion:

$$c(x, t) = c_o \left[\text{erf} \, \frac{x}{(2Dt)^{1/2}} - \text{erf} \, \frac{x}{[4D(t_o + t)]^{1/2}} \right] \qquad 3\text{-}35$$

See figure 3.4 for an illustration of this function, and figure 3.5 is a photo of the Barnacle Electrode prototype.

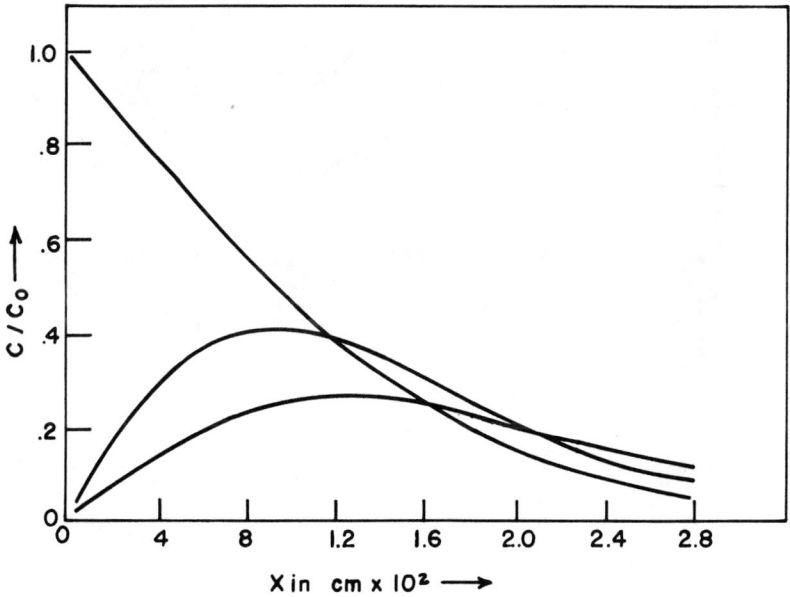

Fig. 3.4.   Rediffusion from the Error Function Complement Profile

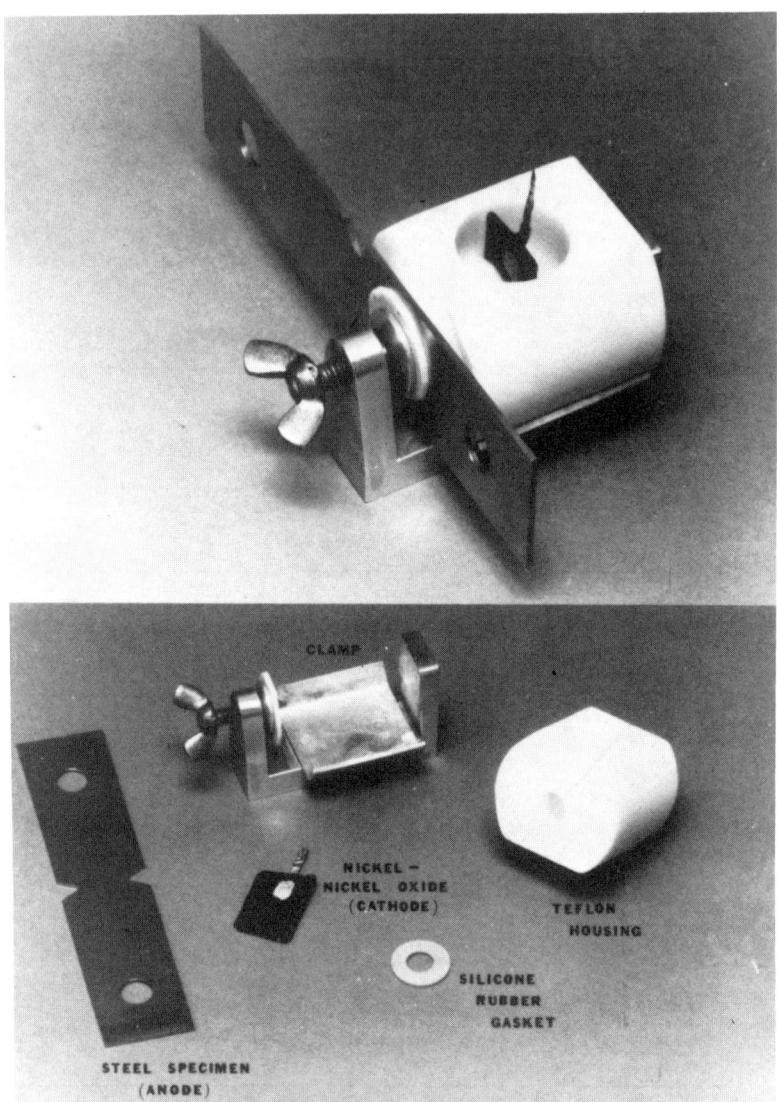

Fig. 3.5. Prototype of the Barnacle Electrode

Source: Copyright ASTM, 1916 Race Street, Philadelphia, PA 19103. Reprinted with permission.

We will now try to explain, as fully as possible, just what is going on at the surface of these diffused specimens.

Thus far, for the surface, we have the equilibrium condition, Eq. A-1. In chapter 2, the equilibrium condition, Eq. 2-3, was replaced by Eq. 2-33, the constant flux condition (Fullenwider, 1976). Here we will look into the exact meaning of this.

What we are looking for is the answer to how the hydrogen gets into the metal in the first place, despite barriers at the surface.

The key boundary condition in the derivation of Eq. A-12 is Eq. A-1.

In diffusion theory, the equilibrium boundary condition, taken from the concept of thermal equilibrium of heat transfer has been used often. It should be noted that Eq. 3-3 contains no information as to how the hydrogen passes the surface and gets into the bulk of the metal.

Rediffusion (solution for concentration profiles and flux out of a precharged specimen) from the error function complement profile is a problem brought to the present author's attention by Dr. Leonard Nanis. The crux to this problem is the boundary condition, 2-33. This is the constant flux boundary condition for what is referred to as chemical rate limitation. This boundary condition is currently finding more applications (Fullenwider, 1975; Early, 1978) than in the past. Here the surface coverage is in equilibrium with the concentration gradient just inside the surface, which immediately changes with any change in coverage. It replaces the equilibrium boundary condition of Eq. 3-3. See figure 3.6 for a potential energy diagram of this case.

Rate limitation, meaning a process slower than the then accepted equilibrium theory, was a term originated by Miller and Smits (1957) early in the 1950s. The type of rate limita-

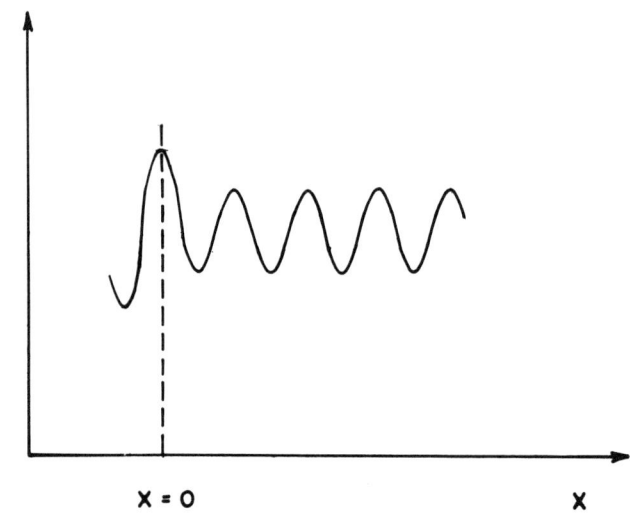

Fig. 3.6. Potential Early Energy Profile for Chemical Rate Limitation

Reprinted by permission of the publisher, The Electrochemical Society, Inc.

tion discussed by them is called "physical rate limitation," and makes use of the radiation boundary condition (Carslaw & Jaeger, 1959). See figure 3.7 where a potential energy profile is illustrated for this situation.

The diagram illustrates diffusion into a semi-infinite solid with a "skin" of thickness δ impeding the motion of the diffusing species into the bulk of the substance covered by the skin. The covering is supposed to be, for example, an oxide of some sort. In figure 3.7, the skin lies between $x = 0$ and $x = -\delta$.

The radiation boundary condition is:

$$[c_e - c(o, t)]K = -D \left[\frac{\partial c(x, t)}{\partial x}\right]_{x = 0} \qquad 3.36$$

BARNACLE ELECTRODE   37

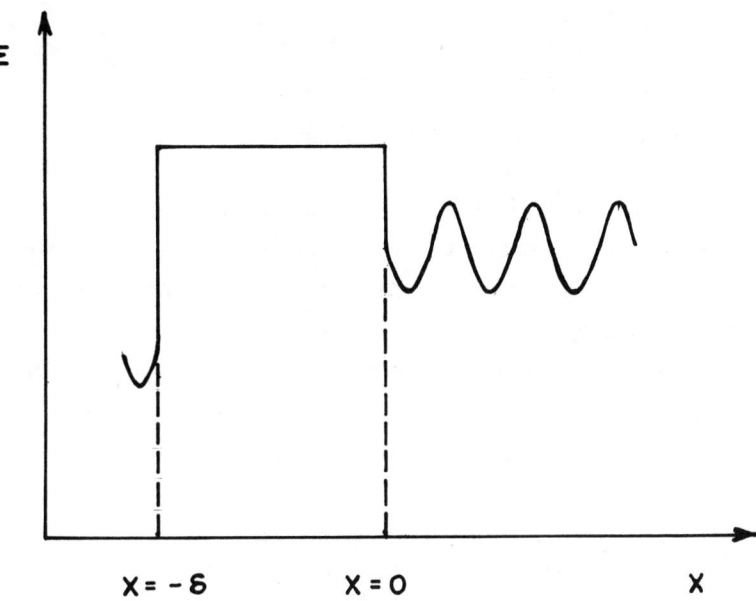

Fig. 3.7. Potential Energy Profile for Physical Rate
Limitation

Reprinted by permission of the publisher, The Electrochemical
Society, Inc.

where the symbols have their usual meaning, c_e is the final
equilibrium concentration at $x = o$ as time becomes infinite,
and K is a constant analogous to an electrical conductance in
that a region of small thickness and small diffusion coefficient may be the equivalent of a large region of larger
diffusion coefficient. A more intimate knowledge of the
system, for example, thickness and diffusivity of the skin,
would allow a better model to be constructed, but this information is usually not available. The solution corresponding
to the error function complement profile is in this case:

$$c(x, t) = c_e \{ \text{erfc} \frac{x}{2(Dt)^{\frac{1}{2}}} - \exp[(Kx + K^2 t)/D]$$

$$\text{erfc } [\frac{x}{2(Dt)^{1/2}} + K \, (\frac{t}{D})^{1/2}] \} \qquad 3\text{-}37$$

There is still another situation concerning oxide films on the surface. Eqs. 3-36 and 3-37 hold only for a "skin" which is permanent.

With the bi-electrode technique, it is possible to have a thin oxide film, particularly in alkaline solutions, which can be imagined to block the surface initially and then slowly come off when hydrogen is introduced, possibly causing the slow initial transient followed by faster reproducible transients as discussed in chapter 2 in the section on charging mode (Bockris, Genshaw, & Subramanyan, 1967).

It was shown (Bockris et al., 1967) that, in the presence of an oxide, the observed diffusion coefficient would have the relation to Δt, the time to remove the oxide film with cathodic charging and $D(I)$, the real diffusion coefficient:

$$D_{obs} = D \, (I) \, \frac{1}{1 + \Delta t / t_{1/2}(I)} \qquad 3\text{-}38$$

where $t_{1/2}(I)$ is the half rise time for the oxide free metal.

At high current densities:

$$\frac{\Delta t}{t_{1/2}(I)} \to 0 \qquad 3\text{-}39$$

Now $t_{1/2}(I)$ varies in proportion to the square of the thickness. Thus, when the membrane thickness is less than a critical value, and oxides are present, $\Delta t/t_{1/2}(I)$ will be finite and cause D to appear to increase with current density.

This mechanism of slow oxide removal explained results with iron (Bockris et al., 1967). The relations seem to offer an explanation for results where the diffusion coefficient appeared to increase with increasing concentration (Namboodhiri & Nanis, 1973), as would happen with Eq. 3-38.

Also, it was found (Namboodhiri & Nanis, 1973) that the effect of the anomalous transients was more pronounced in alkaline solution, where oxides would be more suspect, than in acid solution.

"How general is this explanation" (Early, 1978)? Early finds the slow transient followed by fast transient effect with palladium. Here it seems unlikely that the effect will be due to a slowly removed oxide film. Thus, it appears that in the case of slow removal of hydrogen into the bulk of the metal (palladium), the "slow fast" effect would be due to slowly varying hydrogen coverage, and in the situation of extreme large diffusion coefficient (iron), where hydrogen is taken into the metal rapidly, there is a slowly removable oxide film explanation, at least where they can form (less noble metals).

Getting back to chemical rate limitation and looking for the reason as to how the hydrogen gets into the metal, referring to figure 3.6 for this case, we can see that the potential energy barrier at the surface has the usual chemical shape. The profile in figures 3.6 and 3.7 pertains to endothermic absorption processes; that is, processes in which heat is required to get the hydrogen across the metal interface, the case for many metals, most notably iron. It is not always known how the hydrogen gets into the metal in this situation.

Going into the case of constant flux, j, at input surface for diffusion into a semi-infinite solid, equilibrium between j and surface coverage, for the model needed here, we obtain (see Appendix C):

$$\frac{c(x,t)}{c_o} = e^{-x^2/4Dt} - \frac{x}{2} \left(\frac{\pi}{Dt}\right)^{\frac{1}{2}} \text{erfc} \frac{x}{(4Dt)^{\frac{1}{2}}} \qquad 3\text{-}40$$

There are three possibilities for getting the hydrogen over the potential barrier at the surface:

1. electrochemical overpotential related to high fugacity at the surface and in the bulk of the metal

2. chemical surface reactions
3. hydrogen-phonon interactions at the surface.

Figure 3.8 illustrates the closeness of fit between the equilibrium and constant flux profiles for diffusion into a semi-infinite solid. They can be seen to be very similar once normalized by a factor of 1.7 in the diffusion coefficient, making it difficult to distinguish the two cases. The final, constant flux, concentration profile solution is Eq. 3-40 (Fullenwider, 1974).

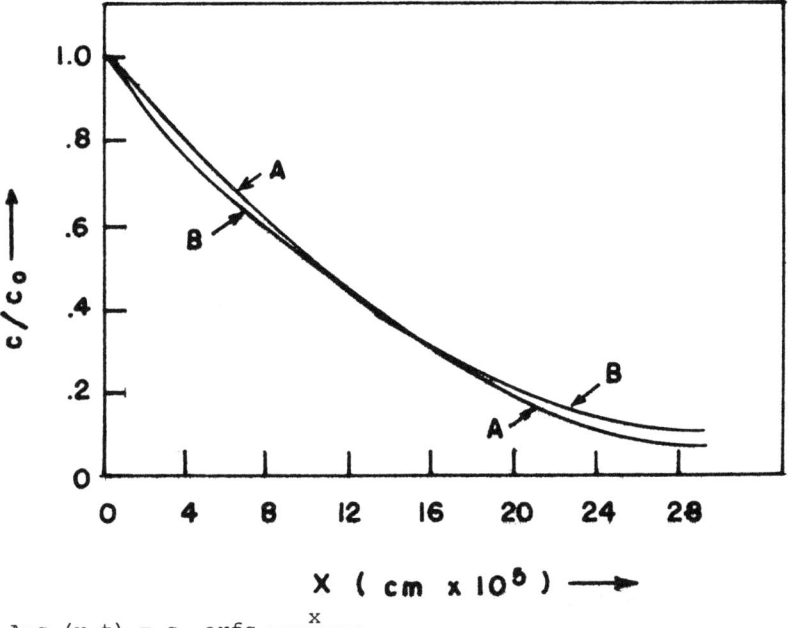

$$A = c(x,t) = c_0 \, \text{erfc} \frac{x}{2 \left(\frac{Dt}{1.7}\right)^{1/2}}$$

$$B = c(x,t) = c_0 \left[e^{-x^2/4Dt} - \frac{x}{2} \left(\frac{\pi}{Dt}\right)^{1/2} \text{erfc} \frac{x}{2(Dt)^{1/2}} \right]$$

Fig. 3.8 Normalized Concentration Profiles

Reprinted by permission of the publisher, The Electrochemical Society, Inc.

Unfortunately, the problem of rediffusion has not been solved for the case of Eq. 3-40. It would be the basis for a useful analytical technique.

Possibility (1) is treated (Bockris & Subramanyan, 1971) for different reaction mechanisms of the hydrogen electrode. The discussion here comes from Bockris and Subramanyan (1971). We are, of course, concerned here only with what is going on at the surface, but embrittlement, as it enters the discussion, is many times the only indication that the hydrogen has indeed passed the surface and entered the bulk of the metal.

In response to a question raised by Dr. Leonard Nanis at a seminar, Bockris and Subramanyan looked into the validity of using the equilibrium thermodynamic relation:

$$e = -\frac{RT}{2F} \ln f_{H_2} \qquad 3\text{-}41$$

in the form:

$$\eta = -\frac{RT}{2F} \ln (f_{H_2})_{int} \qquad 3\text{-}42$$

to express the electrochemical equivalent pressure at the surface and in the bulk of the electrode material, where η is the overpotential associated with hydrogen evolution, and $(f_{H_2})_{int}$ is the fugacity of hydrogen at the surface, in equilibrium with that in cracks or cavities inside the electrode material.

The origin of relation 3-42 is in traditional electrochemistry, in which it is also assumed that the adsorbed H and H+ ions in solution are in equilibrium. It cannot be accepted as a correct relation in the presence of irreversibility in the electron-transfer reaction. In this discussion, the effect of the various mechanisms of hydrogen evolution on the overpotential - $(f_{H_2})_{int}$ relationship was examined in terms of electrochemical kinetics. The assumption of equilibrium between the H at the external surface, the dissolved H, the H at the internal surface of cracks, and H_2 in cracks was retained.

THE ADSORPTION ISOTHERM

The model for the analysis was one in which hydrogen is generated at a constant overpotential on a metal surface, as a result of either an imposed potential or a spontaneous mixed potential as in corrosion.

The prevailing thermodynamic condition is:

$$\mu_{H, \text{ ext surf}} = \mu_{H, \text{ diss}} = \mu_{H, \text{ int surf}} = \mu_{H_2, \text{ crack}} \quad 3\text{-}43$$

where

$$\mu_{H, \text{ ads}} = \mu^o_{H, \text{ ads}} + RT \ln \left(\frac{\Gamma}{1 - \Gamma}\right) \quad 3\text{-}44$$

and

$$\mu_{H_2} = \mu^o_{H_2} + RT \ln f_{H_2} \quad 3\text{-}45$$

Here Γ is the degree of coverage of the H on the electrode surface (external) and f_{H_2} is the fugacity of hydrogen.

From 3-43, 3-44, and 3-45, the following isotherm is obtained:

$$\left(\frac{\Gamma}{1 - \Gamma}\right) = f_{H_2}^{1/2} \exp \left(\frac{(\tfrac{1}{2}\mu^o_{H_2} - \mu^o_{H, \text{ ads}})}{RT}\right) \quad 3\text{-}46$$

Statistically, 3-46 may be expressed as (Moelwyn-Hughes, 1964):

$$\left(\frac{\Gamma}{1-\Gamma}\right) = f_H \left\{\frac{h^3}{(2\pi m_m kT)^{3/2}} \frac{h^2}{8\pi^2 I kT} [1 - \exp(-\frac{h\nu_m}{kT})] \frac{f_{H_2}^{1/2}}{kT}\right\}$$

$$\times \exp \left\{\frac{(-U_H + \tfrac{1}{2} U_m)}{kT}\right\} \quad 3\text{-}47$$

where f_H is the partition function for the adsorbed H atom on the surface, U_H the potential energy of the adsorbed atom, and U_m the potential energy of H_2 molecule in the gas phase. The latter two quantities are related to the heat of desorption ∇H_{des} by

$$\nabla H_{des} = U_m - 2U_H \qquad 3\text{-}48$$

The quantities in 3-47 have their usual meanings. The adsorbed hydrogen atom is assumed to have two degrees of translational freedom and one degree of vibrational freedom perpendicular to the plane of the surface. Therefore, the relation to f_{H_2} becomes

$$\left(\frac{\Gamma}{1-\Gamma}\right) = \left(\frac{2\pi m_H kT}{h^2}\right) \frac{A}{N_S} \frac{kT}{h\mu_H} \left\{ \frac{h^3}{(2\pi m_m kT)^{3/2}} \frac{h^2}{(4\pi^2 IkT)} \right.$$

$$\times \left[1 - \exp\left(-\frac{h\mu_m}{kT}\right)\right] \left.\frac{f_{H_2}}{kT}\right\}^{1/2} \exp\left[\frac{(-U_H + \tfrac{1}{2}U_m)}{kT}\right] \qquad 3\text{-}49$$

where A is the surface area and N_S the total number of sites for adsorption. For the H_2 molecule, $h\mu_m/kT \gg 1$ and for adsorbed atoms, assuming $h\mu_a/kT \gg 1$, 3-49 becomes

$$\left(\frac{\Gamma}{1-\Gamma}\right) = \left(\frac{2\pi m_H kT}{h^2}\right) \frac{A}{N_S} \left[\frac{h^3}{(2\pi m_m kT)^{3/2}} \frac{h^2}{4\pi^2 IkT} \frac{h_{H_2}}{kT}\right]^{1/2}$$

$$\times \exp\left[\frac{(-U_a + \tfrac{1}{2}U_m)}{kT}\right] \qquad 3\text{-}50$$

The quantity U_m is pressure-dependent only to a small extent. Neglecting this, we reduce 3-50 to

$$\left(\frac{\Gamma}{1-\Gamma}\right) = A' f_{H_2}^{1/2} \exp\left(-\frac{U_H}{kT}\right) \qquad 3\text{-}51$$

Taking $U_m = -102$ kcal/mole, the value of A' in 3-51 is 8.9×10^{-40} when f_{H_2} is in atm. From a knowledge of the coverage when $f_{H_2} = 1$ atm or at any known pressure, the value of U_H

can be calculated. In the electrochemical situation, the coverage at 1 atm corresponds to the coverage at the reversible potential (Γ_{rev}), and:

$$\left(\frac{\Gamma_{rev}}{1-\Gamma_{rev}}\right) = A' \exp\left(-\frac{U_H}{kT}\right) \qquad 3\text{-}52$$

or

$$\left(\frac{\Gamma}{1-\Gamma}\right) = \left(\frac{\Gamma_{rev}}{1-\Gamma_{rev}}\right) f_{H_2}^{½} \qquad 3\text{-}53$$

The value of Γ_{rev} on iron, in presence of a solution of pH 4, can be deduced (Bockris & Kita, 1961). It is found to be $\sim 10^{-3}$. Γ_{rev} on Ag in 0.1 M sodium hydroxide solution is found to be $\sim 10^{-2}$ (Devanathan et al., 1959). The calculated value of U_a on Fe for $\Gamma_{rev} = 10^{-3}$ from equation 3-52 is -49.8 kcal/g-atom. In the case of chemisorption of H on Fe from gas phase, the heat of adsorption ΔH_{ads} decreases from -32 kcal to -13 kcal when θ goes from 0 to 1 (Trapnell, 1955). The decrease occurs mainly when θ > 0.5. For platinum in the presence of a solution, a similar decrease of ΔH_{ads} with increase of θ is observed (Breiter, 1961). ΔH_{ads} of H on Pt from the gas phase is -27 kcal/g-mole H_2 while, for adsorption on a surface in NaOH, it is -18 kcal/g-mole H_2 when θ → 0. Thus, here, in the presence of aqueous solution, the heat of adsorption is some 33 percent less than that on a surface in the gas phase (Bockris & Koch, 1961).

The statistical treatment can be used to deduce the value of U_H when the coverage is known.

OVERPOTENTIAL AND INTERNAL PRESSURE OF HYDROGEN IN CAVITIES IN METAL

The relation of η to f_{H_2} (cf. equation 3-43) depends on the mechanism of the hydrogen evolution reaction (her).

INFLUENCE OF MECHANISM TYPE

Fast discharge - slow combination:

$$H^+ + e + M \rightleftarrows M-H, \qquad 3\text{-}54$$

$$M-H + M-H \xrightarrow{rds} 2M + H_2 \qquad 3\text{-}55$$

Expressing the equilibrium condition of Eq. 3-54, we have:

$$k_1 c_{H^+} (1-\Gamma) \exp\left(-\frac{\beta VF}{RT}\right) = k_{-1} \Gamma \exp\frac{(1-\beta)VF}{RT} \qquad 3\text{-}56$$

i.e.:

$$\left(\frac{\Gamma}{1-\Gamma}\right) = \frac{k_1 c_{H^+}}{k_{-1}} \exp\left(-\frac{VF}{RT}\right) \qquad 3\text{-}57$$

or:

$$\left(\frac{\Gamma}{1-\Gamma}\right) = \Gamma_o \exp\left(-\frac{nF}{RT}\right) \qquad 3\text{-}58$$

where:

$$\Gamma_o = \left(\frac{\Gamma_{rev}}{1-\Gamma_{rev}}\right) = \frac{k_1 c_{H^+}}{k_{-1}} \exp\left(-\frac{V_{rev} F}{RT}\right) \qquad 3\text{-}59$$

From 3-52, 3-57, and 3-58,

$$f_{H_2} = \exp\left(-\frac{2\eta F}{RT}\right) \qquad 3\text{-}60$$

The pressure calculated upon the assumption of the classical Nernst equation would be as in 3-60; the Nernst pressure will be obtained in cavities, if the other stated assumptions are applicable. Of all mechanisms of the her, this is one of the most embrittling (i.e., gives the highest f_{H_2}, for a given η).

Slow discharge - fast combination:

$$H^+ + e + M \xrightarrow{rds} M-H, \qquad 3\text{-}61$$

$$M-H + M-H \rightleftarrows 2M + H_2 \qquad 3\text{-}62$$

For Eq. 3-62, one obtains:

$$\left(\frac{\Gamma}{1-\Gamma}\right) = \frac{k_{-2}}{k_2} p_{H_2} \qquad 3\text{-}63$$

Thus, Γ is independent of potential and p_{H_2} is normally equal to 1 atm. Hence, Γ is always approximately equal to the coverage at the reversible potential. This is an idealization. In fact, there will be a small increase of Γ over Γ_{rev} and f_{H_2} will be somewhat greater than 1 atm, but will never increase to magnitudes that will damage the metal. This mechanism will not contribute to embrittlement by pressure.

Slow discharge - fast electrochemical desorption:

$$H^+ + e + M \xrightarrow{rds} M - H \qquad 3\text{-}64$$

$$M - H + H^+ + e \rightleftarrows M + H_2 \qquad 3\text{-}65$$

For Eq. 3-65, one obtains

$$k_2 c_{H^+} \Gamma \exp\left(-\frac{\beta VF}{RT}\right) = k_2 (1-\Gamma) p_{H_2} \exp\left[(1-\beta)\frac{VF}{RT}\right] \qquad 3\text{-}66$$

Hence,

$$\left(\frac{\Gamma}{1-\Gamma}\right) = \frac{k_{-2}}{k_2 c_{H^+}} p_{H_2} \exp\left(\frac{VF}{RT}\right) \qquad 3\text{-}67$$

In this case, p_H on the r.h.s. of 3-67 is constant, at 1 atm. Therefore,

$$\left(\frac{\Gamma}{1-\Gamma}\right) = \Gamma_o \exp\left(\frac{\eta F}{RT}\right) \qquad 3\text{-}68$$

where

$$\Gamma_o = \left(\frac{\Gamma_{rev}}{1-\Gamma_{rev}}\right) = \frac{k_2}{k_2 c_{H^+}} p_{H_2} \exp\left(\frac{V_{rev} F}{RT}\right) \qquad 3\text{-}69$$

Now relating η and f_{H_2} through the isotherm, we have

$$f_{H_2} = \exp\left(\frac{2\eta F}{RT}\right) \qquad 3\text{-}70$$

Thus (as η is negative), the pressure developed in the cavities will be smaller than 1 atm. This the least embrittling mechanism, for the surface concentration of H is less than that which exists in equilibrium with H_2 at 1 atm.

Fast discharge - slow electrochemical desorption:

$$H^+ + e + M \rightleftarrows M - H \qquad \qquad 3\text{-}71$$

$$H^+ + e + M - H \xrightarrow{rds} M + H_2 \qquad \qquad 3\text{-}72$$

For reaction 3-71:

$$k_1 c_{H^+} (1-\Gamma) \exp\left(-\frac{\beta VF}{RT}\right) = k_{-1} \Gamma \exp\left\{(1-\beta)\frac{VF}{RT}\right\} \qquad 3\text{-}73$$

Hence

$$\left(\frac{\Gamma}{1-\Gamma}\right) = \frac{k_1 c_{H^+}}{k_{-1}} \exp\left(-\frac{VF}{RT}\right) \qquad \qquad 3\text{-}74$$

or

$$\left(\frac{\Gamma}{1-\Gamma}\right) = \Gamma_o \exp\left(-\frac{\eta F}{RT}\right) \qquad \qquad 3\text{-}75$$

where

$$\Gamma_o = \left(\frac{\Gamma_{rev}}{1-\Gamma_{rev}}\right) = \frac{k_1 c_{H^+}}{k_{-1}} \exp\left(-\frac{V_{rev} F}{RT}\right) \qquad 3\text{-}76$$

From 3-53 and 3-74, we obtain:

$$f_{H_2} = \exp\left(-\frac{2\eta F}{RT}\right) \qquad \qquad 3\text{-}77$$

In this case, the pressure developed in cavities will be the same as predicted from the Nernst equation.

Coupled discharge - combination:

$$H^+ + e + M \xrightarrow{rds} M - H \qquad \qquad 3\text{-}78$$

$$M - H + M - H \xrightarrow{fast} 2M + H_2 \qquad \qquad 3\text{-}79$$

At any potential other than V_{rev} (reversible potential), the following relationship holds:

$$2k_1 c_{H^+} (1-\Gamma) \exp\left(-\frac{\beta VF}{RT}\right) = k_2 \Gamma^2 \qquad \qquad 3\text{-}80$$

or

$$\frac{\Gamma^2}{(1-\Gamma)} = \frac{2k_1 c_{H^+} \exp\left(-\frac{\beta VF}{RT}\right)}{k_2} \qquad 3\text{-}81$$

Case 1. When $\Gamma \ll 1$:

$$\frac{\Gamma^2}{(1-\Gamma)} \sim \frac{\Gamma^2}{(1-\Gamma)^2} = \frac{2k_1 c_{H^+} \exp\left(-\frac{\alpha VF}{RT}\right)}{k_2} \qquad 3\text{-}82$$

The equilibrium fugacity of H_2 in the cavity is given by:

$$\left(\frac{\Gamma}{1-\Gamma}\right)^2 \left(\frac{1-\Gamma_{rev}}{\Gamma_{rev}}\right)^2 = f_{H_2} \qquad 3\text{-}83$$

At the equilibrium potential, since the forward reaction rates are no longer coupled, one cannot obtain Γ_{rev} from 3-82 by letting $V \rightarrow V_{rev}$. However, $(\Gamma_{rev}/1-\Gamma_{rev})$ can be obtained by considering the equilibrium state of either Eqs. 3-78 or 3-79. As an example from Eq. 3-79,

$$k_2 \Gamma_{rev}^2 = k_{-2}(1-\Gamma_{rev})^2 p_{H_2} \qquad 3\text{-}84$$

Hence:

$$\left(\frac{\Gamma_{rev}}{1-\Gamma_{rev}}\right)^2 = \frac{k_{-2} p_{H_2}}{k_2} \qquad 3\text{-}85$$

From 3-82, 3-83 and 3-84,

$$f_{H_2} = \frac{2k_1 c_{H^+} \exp\left(-\frac{\beta VF}{RT}\right)}{k_{-2} p_{H_2}} \qquad 3\text{-}86$$

or:

$$f_{H_2} = \frac{2k_1 c_{H^+} \exp\left(-\frac{\beta V_{rev} F}{RT}\right) \exp\left(-\frac{\beta \eta F}{RT}\right)}{k_{-2} p_{H_2}} \qquad 3\text{-}87$$

The quantity p_{H_2} represents the atmospheric pressure.

Thus, for the coupled discharge-combination mechanism, the internal hydrogen fugacity f_{H_2} cannot be obtained without

BARNACLE ELECTRODE 49

a knowledge of various constants. By making a reasonable approximation, it is possible to calculate the value of the internal fugacity in terms of overpotential. For example, one assumes that the discharge step is in equilibrium up to an overpotential of $\eta = 2.303\ RT/F$. From Eq. 3-78,

$$k_1 c_{H^+} (1 - \Gamma) \exp\left(-\frac{\beta VF}{RT}\right) = k_{-1} \Gamma \exp\left\{(1-\beta)\frac{VF}{RT}\right\} \qquad 3\text{-}88$$

Therefore

$$\left(\frac{\Gamma_{lim}}{1-\Gamma_{lim}}\right)^2 = \left(\frac{\Gamma_{rev}}{1-\Gamma_{rev}}\right)^2 \exp\left(-\frac{2\eta F}{RT}\right) \qquad 3\text{-}89$$

where $\Gamma_{lim} = \Gamma$ for $\eta < 2.303\ RT/F$.

From 3-82 for $\Gamma \ll 1$,

$$\frac{\Gamma^2}{(1-\Gamma)^2} = \frac{2k_1 c_{H^+} \exp\left(-\frac{\beta V_R F}{RT}\right)}{k_2} \exp\left(-\frac{\beta \eta F}{RT}\right) =$$

$$K_0 \exp\left(-\frac{\beta \eta F}{RT}\right) \qquad 3\text{-}90$$

At $\eta - 2.303\ RT/F$, 3-89 and 3-90 are both applicable. Hence:

$$K_0 = \left(\frac{\Gamma_{rev}}{1 - \Gamma_{rev}}\right)^2 \times 10^{1.5} \qquad 3\text{-}91$$

Substituting for K_0 from 3-91 into 3-90:

$$\frac{\Gamma^2}{(1 - \Gamma)^2} = \left(\frac{\Gamma_{rev}}{1 - \Gamma_{rev}}\right)^2 10^{1.5} \exp\left(-\frac{\alpha \eta F}{RT}\right) \qquad 3\text{-}92$$

Hence, f_{H_2} when $\eta > 2.3\ RT/F$ is given by:

$$f_{H_2} = 10^{1.5} \exp\left(-\frac{\eta F}{2RT}\right) \qquad 3\text{-}93$$

when $\alpha = 0.5$.

Thus, the pressure developed is less than the Nernst pressure.

Coupled discharge - electrochemical desorption:

$$H^+ + e + M \xrightarrow{fast} M-H \qquad 3\text{-}94$$

$$M-H + H^+ + e \xrightarrow{rds} M + H_2 \qquad 3\text{-}95$$

From Eqs. 3-94 and 3-95

$$k_1 (1-\Gamma) c_{H^+} \exp\left(-\frac{\beta VF}{RT}\right) = k_2 \Gamma c_{H^+} \exp\left(-\frac{\beta VF}{RT}\right) \qquad 3\text{-}96$$

The above equation, however, does not hold good at the reversible potential. At potentials sufficiently large to allow the coupled condition for Eqs. 3-94 and 3-95:

$$\left(\frac{\Gamma}{1-\Gamma}\right) = \frac{k_1}{k_2} \qquad 3\text{-}97$$

Now, considering the equilibrium conditions of Eq. 3-94 and using the isotherm given by 3-53:

$$f_{H_2} = \frac{k_{-1}}{k_2^2 c_{H^+}^2} \exp\left(\frac{2V_{rev} F}{RT}\right) \qquad 3\text{-}98$$

The coverage is independent of potential. In this case, as in the previous one, an estimation of the internal pressure requires a knowledge of the constants. However, for the present mechanism, $\Gamma/(1-\Gamma)$ is usually found not to exceed unity. From 3-51, using the value (relevant for Fe) of U_a as -50 kcal/g-atom, the maximum pressure is calculated as 4.4×10^{-7} atm. Therefore, this mechanism is powerful enough to cause embrittlement in metals due to hydrogen pressure.

A more accurate way of determining the internal fugacity would be to obtain $[(\Gamma_{rev}/(1-\Gamma_{rev})]$ from the preceding stage of the reaction when we have fast discharge-slow electrochemical desorption (non-coupled) mechanism prevailing. For this mechanism, $d\eta/d \ln i$ is known to be equal to $2RT/3F$. Beyond this region, we have $d\eta/d \ln i = 2RT/F$. At the point of intersection of the two Tafel slopes, the internal fugacity

BARNACLE ELECTRODE

f_{H_2} must be the same. From equations 3-75 and 3-76, we obtain:

$$\left(\frac{\Gamma}{1-\Gamma}\right) = \left(\frac{\Gamma_{rev}}{1-\Gamma_{rev}}\right) \exp\left(-\frac{\eta F}{RT}\right) \qquad 3\text{-}99$$

Hence, at the potential of intersection of the Tafel slope (η^*), from 3-97 and 3-99, we get:

$$\left(\frac{\Gamma}{1-\Gamma}\right)^2 = \left(\frac{k_1}{k_2}\right)^2 = \left(\frac{\Gamma_{rev}}{1-\Gamma_{rev}}\right)^2 \exp\left(-\frac{2\eta^* F}{RT}\right) \qquad 3\text{-}100$$

Hence, the internal fugacity f_{H_2} is given by:

$$f_{H_2} = \exp\left(-\frac{2\eta^* F}{RT}\right) \qquad 3\text{-}101$$

which shows that f_{H_2} for the mechanism of coupled discharge (fast) and slow electrochemical desorption is a constant and independent of overpotential. It can only be estimated if η^* is known experimentally, i.e., if there is evidence concerning the changeover for $d\eta/d \ln i$. η^* represents the potential at which fast discharge-electrochemical desorption changes to coupled discharge-electrochemical desorption.

It was also found (Bockris & Subramanyan, 1971) that equivalent pressures would be much less than those generally assumed, e.g., a fugacity of 10^6 atmospheres would result in a pressure of only about 10^4 atmospheres.

Therefore, it can be seen that not all mechanisms produce embrittlement. There are, in addition to the cases reported here, "no overpotential" situations where dry metal is exposed to hydrogen, resulting in embrittlement (Johnson, 1969).

Going back to the possibility of pp. 39f. the first approach for getting hydrogen over the potential barrier seems likely at first thought in most of these cases.

Possibility (1) appears acceptable for most instances where there is an appreciable cathodic overpotential present at the surface. There are still, however, instances where embrittlement occurs without high overpotentials (Johnson, 1969).

The second possibility can be ruled out only in the circumstances in which experimental data is available. In the case of undervoltage experiments using potentiostatic charging with the hydrogen palladium system (Fullenwider, 1976), it has been shown that infinite surface source conditions, that is, the situation where the diffusing source at the surface does not become depleted during the experiment, rather, instances where the equilibrium condition would be expected to apply, resulted in the constant flux type boundary condition.

We are left then with a constant flux situation except where there is a significant overpotential. This may be explained as being due to a constant flux to and from the surface, of phonons, interacting with the protons of a variable Γ in the form of apertures at the surface of the metal creating a "pumping" action (constant flux) there. It is not known how general this phenomenon is, but it seems to explain the present data (Fullenwider, 1976; Early, 1978) concerning the "no overpotential" situations.

4
Statistical Studies

INTRODUCTION

In this chapter, the treatment of which will be useful in the discussion of the last four chapters, we take two cases: (1) that of a three-dimensional gas of hydrogen molecules outside the metal phase in equilibrium with a gas of partially shielded protons inside the metal. This is known as "external hydrogen" by present workers (Thompson & Bernstein, 1975). And (2) equilibrium between hydrogen in the form of two-dimensional gas at a surface, in equilibrium with a three-dimensional gas of protons inside the metal, known as "internal hydrogen."

The treatment will involve grand partition functions. The discussion of external hydrogen will follow the usual treatment (Fowler & Guggenheim, 1965; Lacher, 1937). That of internal hydrogen will be based on the same concepts but is more recent work (Fullenwider, 1974).

We will say a few words about grand partition functions and why they are used here. With ordinary partition functions, it is customary to start with a given energy, E, and arrive at a temperature $\theta' = e^{-1/kT}$ (Fowler & Guggenheim, 1965), a definition, while particle number, N, is fixed. The absolute activity, $\lambda = e^{\mu/kT}$, another definition (Fowler & Guggenheim, 1965), is determined by the N's. In other words, the temperature dictates the energy. The grand functions are obtained by summing:

$$\lambda_A^{N_A} \lambda_B^{N_B} \ldots \ldots \theta'^E \qquad 4-1$$

over all possible N's and all accessible states, making the λ's determine the N's, i.e., here the concentration will be allowed to vary as a function of the absolute activity. Concentrations in metal-hydrogen systems vary considerably.

Physically, this amounts to, instead of considering completely isolated systems with fixed N, taking one in contact with another very large system with which it may exchange particles and energy. We then determine the properties of our system in equilibrium with the large system. This is best done with grand partition functions.

GRAND PARTITION FUNCTION FOR METAL-HYDROGEN SYSTEMS

We proceed by picking out of the discussion (Fowler & Guggenheim 1965) those matters which apply to metal-hydrogen systems. The standard state of these systems is chosen as that of the pure metal crystal with no hydrogen present. The pure crystal has a partition function $K(T)$, and it is taken as an approximation that any changes, say the introduction of hydrogen atoms at interstitial positions, merely changes the energy of the system by a constant amount but has no effect on the number or vibration frequencies of the states. This new configuration has a general partition function:

$$K(T) e^{-W/kT} \qquad 4-2$$

where W is the energy by which the new configuration exceeds the energy of the standard perfect metal configuration.

STATISTICAL STUDIES

For metal-hydrogen systems, introducing:

$$\lambda_H = e^{(\mu/kT)} \qquad 4\text{-}3$$

the absolute activity, and three-dimensional, translational ideal gas partition function:

$$f'_H(T) = (2\pi m_H kT)^{3/2} V/h^3 \qquad 4\text{-}4$$

where V is the volume of the metal-hydrogen system, and m_H is the mass of a proton, and the other symbols have their meanings (see nomenclature), we arrive at the expression for the total grand partition function factor for metal-hydrogen systems:

$$\Gamma'_H(T,\lambda_H) = \sum_{N_H} \{[(\alpha N)!/N_H!(\alpha N - N_H)!] \lambda_H f'_H(T)\rho\}^{N_H}$$

$$\times \exp[-(N_H w_H + \tfrac{1}{2} N_H^2 w_{HH}/\alpha N)/kT] \qquad 4\text{-}5$$

where $(\alpha N)!/N_H!(\alpha N-N_H)!$ is the number of ways N_H protons can be put on αN available sites and where αN is the number of interstitial sites, N_H is the total number of hydrogen atoms distributed among the αN sites in the metal, ρ is the nuclear spin weight which can be taken as unity in the absence of a magnetic field, w_H is the potential energy of the protons relative to the state of infinite dispersion outside the metal, and w_{HH} is a constant representing proton-proton interactions taken here to be proportional to concentration.

In this approximation, it is assumed that the protons are free to move in a uniform potential energy w_H, relative to the state of infinite dispersion of H atoms over a volume V_H, roughly the volume, V, of the metal.

These relations are all for hydrogen dissolved in the metal. All we need to do here to get expressions for "external hydrogen" is to develop a relation for the three-dimensional gas outside the metal and eliminate the λ's for the two cases, dilute and more concentrated.

We go now on with the calculation of the most probable states. In this approximation (that of a dilute system), we assume that proton-proton interactions are negligible. Thus:

$$w_{HH} = 0, \qquad 4\text{-}6$$

and the number of interstitial protons is assumed to be much less than the number of interstitial sites available:

$$N_H \ll \alpha N \qquad 4\text{-}7$$

and we have for the grand partition function, from Eq. 4-5:

$$\Gamma_H'(T,\lambda_H) = \sum_{N_H} [(1/N_H!)\lambda_H f_H'(T)\rho]^{N_H} \exp(-N_H w_H/kT) \qquad 4\text{-}8$$

Taking logarithms of both sides of Eq. 4-8 and setting the derivative of the general term with respect to N_H equal to zero, that is, replacing the series by its maximum term:

$$\partial \ln [\text{term of sum: } [\Gamma_H'(T,\lambda_H)]/\partial N_H = 0 \qquad 4\text{-}9$$

we arrive at the expression for the greatest term of the series (most probable state):

$$N_H = \lambda_H f_H'(T) \rho \exp(-w_H/kT) \qquad 4\text{-}10$$

For more densely populated systems, higher concentrations, we use Eq. 4-5, as is, and obtain for the most probable state in this case:

$$\frac{N_H}{\alpha N - N_H} = \lambda_H f_H'(T)\rho \exp[-(w_H + N_H w_{HH}/\alpha N)/kT] \qquad 4\text{-}11$$

The step of introducing equilibrium between the small and large systems has not yet been accomplished. The most probable state, however, has been established. To get equilibrium

STATISTICAL STUDIES

expressions for two systems, all that is necessary is to set the λ's for the two systems equal to each other.

External Hydrogen: Dilute Case

External hydrogen is simply hydrogen that "starts out" on the outside of the metal. Here we use Eq. 4-8, and we need an expression for λ_H, the absolute activity of a gas of hydrogen molecules at pressure p. It is given by (Fowler & Guggenheim, 1965):

$$\lambda_H = \lambda_{H_2}^{1/2} (\frac{p}{kT})^{1/2} e^{-\frac{1}{2}\chi_d/kT} \times \{\frac{2\pi 2m_H kT}{h^3} - \frac{8\pi^2 IkT}{h^2} \frac{p}{2}\}^{-1/2} \qquad 4\text{-}12$$

where χ_d is the energy diffusion between the molecule in its lowest state and two free H atoms in their lowest states (not to be confused with the energy required to break the molecule bond), and I is the moment of inertia of the H_2 molecule. Combining Eqs. 4-10 and 4-12, that is, eliminating λ_H, we have:

$$\frac{N_H}{V_H} = (\frac{m_H^3}{16\pi I^2 h^2 kT})^{1/4} p^{1/2} e^{-(w_H + \frac{1}{2}\chi_d)/kT} \qquad 4\text{-}13$$

From Eq. 4-13, it can be seen that the concentration N_H/V_H will depend exponentially on χ_d, a positive quantity.

For metals such as iron and platinum, which are endothermic occluders, the solubility will be small in this approximation, since both χ_d and w_H will be positive. Also, metals such as titanium, which are covered by oxides almost impenetrable to hydrogen, equilibrium would probably never be established, unless the oxide is removed by, for example, bombarding with ions or exposure to hydrofluoric acid. There was an interesting experiment done (DeLuccia, 1975), coating oxide-free Ti with Pd, and determining a diffusion coefficient of 5.6 x 10^{-7} $cm^2 sec^{-1}$ with the bi-electrode technique. The diffusion coefficient had previously been thought by some investigators (Bockris et al., 1970) to be in the range of $10^{-16} cm^2 sec^{-1}$, following from experiments with the barnacle electrode, so the oxide can be seen to make quite a difference.

In Eq. 4-13, the quantity w_H is negative for exothermic occluders and positive for endothermic. Thus, solubilities will be greater in the exothermic case.

External Hydrogen: More Concentrated Case

We obtain an expression for the absolute activity in a more concentrated case of external hydrogen (Lacher, 1937):

$$\lambda_H = \frac{\theta''}{1-\theta''} \frac{e^{(w_H + 2\theta'' w_{HH})/kT}}{\rho q_H(T)} \qquad 4\text{-}14$$

where θ'' (Lacher, 1937) is the fraction of sites occupied by hydrogen atoms, $N_H/\alpha N$, $q_H(T)$ is the vibrational partition function of an absorbed H atom far removed from other H atoms, which we may set equal to 1.

Setting λ_H of 4-14 equal to λ_H of Eq. 4-12, we get for the equilibrium of hydrogen between the metal and the gas:

$$\frac{\theta''}{1-\theta''} = \left(\frac{p}{kT}\right)^{\frac{1}{2}} e^{-(w_H + 2\theta'' w_{HH} + \frac{1}{2}\chi_d)/kT}$$

$$\times \left\{ \frac{(2\pi 2m_H kT)^{3/2}}{h^3} \frac{8\pi^2 AkT}{2h^2} \right\}^{\frac{1}{2}} \qquad 4\text{-}15$$

This relation is analogues to Langmuir's isotherm.

It can be seen that:

$$p \propto [\theta''/(1 - \theta'')]^2 \qquad 4\text{-}16$$

and as θ'' approaches unity, greater external pressures will be required to introduce more hydrogen.

STATISTICAL STUDIES 59

Internal Hydrogen: Dilute Case

In the example of internal hydrogen (Fullenwider, 1974), it is assumed that hydrogen is already present in the dissociated form inside the bulk of the metal, due to, e.g., hydrogen dissociation and entry at the surface, high equivalent pressure electrochemical mechanism such as in electroplating, pumping action of rate limitation, etc. This treatment is similar to that of external hydrogen with the exception that it is not necessary for the hydrogen to cross any external surfaces, and at the surface of cracks and voids in the bulk of the metal, we need only expressions for the spreading pressure of protons adsorbed there. Here also χ_d will be absent from the expressions.

First, we must define some quantities. From Eq. 4-4:

$$f_H' (T) = (2\pi m_H kT)^{3/2} V/h^3 \qquad 4\text{-}17$$

for the three-dimensional ideal gas as appears in the expression for the grand partition function, and:

$$f_H (T) = 2\pi m_H kT \, A/h^2 \qquad 4\text{-}18$$

the partition function for an ideal two-dimensional gas. Here, A is the surface area. ϕ, the spreading pressure is given:

$$\phi = -\partial F^{ads}/\partial A \qquad 4\text{-}19$$

where F^{ads} is the Gibbs free energy of adsorption of a species. F^{ads} for an ideal layer of hydrogen adsorbed on a surface is:

$$F_H^{ads} = -kT\lambda_H f_H (T) + N_H kT \ln \lambda_H \qquad 4\text{-}20$$

From the discussion of the most probable state, we have:

$$N_H = \lambda_H f_H' (T) \, \rho \, \exp(-w_H/kT) \qquad 4\text{-}21$$

and from Eqs. 4-17, 4-18, 4-19 and 4-20:

$$F_H^{ads} = -kT \left[\frac{N_H h^3 \exp(w_H/kT)}{(2\pi m_H kT)^{3/2} \rho V}\right] \frac{2\pi m_H kT}{h^2} A$$

$$+ N_H kT \ln \left[\frac{N_H h^3 \exp(w_H/kT)}{(2\pi m_H kT)^{3/2} \rho V}\right] \quad 4\text{-}22$$

and assuming a cubic symmetry:

$$A = V^{2/3} \quad 4\text{-}23$$

we get for ϕ, from Eqs. 4-19 and 4-22:

$$\phi = \frac{3}{2} N_H, \text{surface } kT/A - \frac{N_{H,bulk} kTh \exp(w_H/kT)}{2 V \rho (2\pi m_H kT)^{1/2}} \quad 4\text{-}24$$

Eq. 4-24 relates ϕ to a combination of surface and bulk effects for a dilute gas inside the metal. The surface term can be for either an internal surface such as that found at a crack or void, or the external surface of the metal.

Internal Hydrogen: More Concentrated Systems

Proceeding similarly as in the last section, we get as the free energy of absorption:

$$F_H^{ads} = -kT \left(\frac{N_H h^3 \exp[(w_H + N_H w_{HH}/\alpha N)/kT]}{(\alpha N - N_H)(2\pi m_H kT)^{3/2} \rho V}\right)$$

$$\times \frac{2\pi m_H kT}{h^2} A + N_H kT \quad 4\text{-}25$$

and with Eq. 4-19 for ϕ and Eq. 4-23:

$$\phi = \frac{3}{2} N_{H,surface} kT/A - \frac{N_H kTh \exp[w_H + N_H w_{HH}/\alpha N)/kT]}{(\alpha N - N_H) 2 V \rho (2\pi m_H kT)^{1/2}} \quad 4\text{-}26$$

As in the discussion of external hydrogen, taking notice (Lacher's notation):

STATISTICAL STUDIES

$$\frac{N_H}{\alpha N - N_H} = \frac{\theta''}{1 - \theta''} \qquad 4\text{-}27$$

the spreading pressure will diverge to large values as $\theta'' \to 1$.

SUMMARY

In summary, we have here three cases: external hydrogen, hydrogen "starting out" external to the metal surface, and hydrogen "starting out" internal to the surface. The first and third are dilute, and the second and fourth more concentrated.

5
The Hydrogen Electrode

INTRODUCTION

The hydrogen electrode is a good example of external hydrogen. With the normal hydrogen electrode (NHE), hydrogen gas is bubbled about a metal at a pressure of one atmosphere, the pH being at 0. In the instance of the reversible hydrogen (RHE), the pH is allowed to vary. This electrode is useful as a reference. The potential varies 59 mV per unit pH value at 25° C.

It is obvious from the text, so far, that hydrogen will enter the metal of the electrode, systems varying considerably from metal to metal. The significance of this hydrogen dissolved in the metal (now internal hydrogen) will be brought out at the end of the chapter. First, we shall concern ourselves with the mechanism of the hydrogen evolution reaction (HER) on various metals. This type of work has been developed to a high degree (McBreen & Genshaw, 1967). The effect of internal hydrogen on the RHE will follow (Fullenwider, 1974).

THE HYDROGEN ELECTRODE

REACTION MECHANISMS

Overall Reaction

The overall reaction for hydrogen evolution, the determination of which is the first step in any mechanism study, is different in acid and alkaline solution. We have, in acid solution, the following overall reaction:

$$2 H_3O^+ + 2e^- \rightarrow H_2 + 2H_2O \qquad 5\text{-}1$$

and in alkaline solution:

$$2 H_2O + 2e^- \rightarrow H_2 + 2OH^- \qquad 5\text{-}2$$

Reaction Paths

There are two reaction paths which are most accepted. The atomic hydrogen desorption path in acid solution:

$$H_3O^+ + e^- + M \rightarrow MH + H_2O \qquad 5\text{-}3$$

$$2MH \rightarrow H_2 + 2M \qquad 5\text{-}4$$

and in alkaline solution:

$$H_2O + e^- + M \rightarrow MH + OH^- \qquad 5\text{-}5$$

$$2MH \rightarrow H_2 + 2M \qquad 5\text{-}6$$

where MH is the adsorbed hydrogen atom on the metal, M.

The other path is the electrochemical desorption path. In acid solution:

$$H_3O^+ + e^- + M \rightarrow MH + H_2O \qquad 5\text{-}7$$

$$H_3O^+ + e^- + MH \rightarrow H_2 + M + OH^- \qquad 5\text{-}8$$

and in alkaline solution:

$$H_2O + e^- + M \rightarrow MH + OH^- \qquad 5\text{-}9$$

$$H_2O + e^- + MH \rightarrow H_2 + M + OH^- \qquad 5\text{-}10$$

Current Potential Relations

Taking as an example the discharge reaction which is the first step in the atomic desorption path, Eq. 5-3, we get for the velocity (a factor of two in some of the rate expressions has been corrected):

$$i = Fka_{H_3O^+} \times (1 - \Gamma) e^{\frac{-W}{RT}} \cdot e^{-\beta(V_E - V_{rev}) F/RT} \qquad 5\text{-}11$$

where $a_{H_3O^+}$ is the activity of hydronium ions at the electrode surface, Γ is the hydrogen coverage of the metal, β is ~ 0.5, a constant, V_E the potential of the electrode, V_{rev} the reversible hydrogen potential, $V_E - V_{rev} = \eta$, the overpotential, and F is the Faraday.

When the reaction, Eq. 5-3, takes place close to the reversible potential, the reverse of Eq. 5-3 must be taken into consideration. We get in this situation for the rate:

$$i = F[k_1 a_{H_3O^+} (1-\lambda) e^{-\beta\eta F/RT} - k_{-1} \Gamma e^{(1-\beta)\eta F/RT}] \qquad 5\text{-}12$$

where k_1 and k_{-1} are the forward and reverse rate constants, respectively.

At the reversible potential, no net current flows, but the forward and reverse reactions are balanced. We have:

$$\vec{i} = \overleftarrow{i} = i_o \qquad 5\text{-}13$$

The exchange current, i_o, is an important constant, say, if we are less than 50 mV away from the reversible potential, it is possible to write:

$$i = i_o e^{-\beta\eta F/RT} \qquad 5\text{-}14$$

THE HYDROGEN ELECTRODE

or:

$$\eta = \frac{RT}{\beta F} \ln i_o - \frac{RT}{\beta F} \ln i \qquad 5\text{-}15$$

or the familiar Tafel equation:

$$\eta = a + b \log i \qquad 5\text{-}16$$

Some values of i_o are listed in table 5.1.

Table 5.1. Exchange Current Densities for Various Metals

Metal	i_o (A cm^{-2})
Hg	10^{-13}
Fe	10^{-7}
Ni	10^{-7}
Pd	10^{-3}
Pt	5×10^{-3}

Hg is a high overpotential metal, Fe and Ni intermediate, and Pd and Pt low.

Source: J. McBreen and M.A. Genshaw, <u>Stress Corrosion Cracking</u>, Proceedings of a conference at Ohio State University, Columbus, Ohio, September 1967. Reprinted by permission of the National Association of Corrosion Engineers.

We must, now, discuss adsorption isotherms. For small Γ, the Langmuir isotherm (non-interacting species) holds. Equating the rates of adsorption and desorption:

$$k_F (1-\Gamma) \exp(-\beta \frac{\eta F}{RT}) = k_b \Gamma \exp(1-\beta) \frac{\eta F}{RT} \qquad 5\text{-}17$$

where k_1 and k_{-1} are constants, Γ the fraction of surface covered, $\beta \cong \frac{1}{2}$, and the other constants have their usual meaning (see nomenclature).

$$\frac{\Gamma}{1 - \Gamma} = k \exp\left(\frac{\eta F}{RT}\right) \qquad 5\text{-}18$$

At intermediate coverages, there is interaction between particles on the electrode surface. Here we introduce the Tempkin isotherm. For the free energy of adsorption, we have:

$$\Delta F_{ads} = r\, \Gamma_{ads} + \text{constant} \qquad 5\text{-}19$$

The isotherm is:

$$r\, \Gamma + RT \ln \frac{\Gamma}{1 - \Gamma} = -F + \text{constant} \qquad 5\text{-}20$$

or:

$$\Gamma \sim \text{constant} - \frac{\eta F}{r} \qquad 5\text{-}21$$

Now, let us get back to the current potential relations.

SLOW DISCHARGE - FAST RECOMBINATION, LANGMUIR

Writing the rate as:

$$i_1 = F k_1 c_{H+} (1 - \Gamma)\, e^{-\beta \eta F / RT} \qquad 5\text{-}22$$

To get Γ, it is assumed that the catalytic step is in equilibrium:

$$k_2 (k'\Gamma)^2 = k_{-2} (1 - \Gamma)^2\, e^{-\beta \eta F / RT} \qquad 5\text{-}23$$

and:

$$\frac{\Gamma}{1 - \Gamma} = \frac{k_{-2}^{\frac{1}{2}}}{k_2^{\frac{1}{2}}} \cdot \frac{p^{\frac{1}{2}}}{k'} \qquad 5\text{-}24$$

where p is the hydrogen partial pressure. Some algebra gives:

$$\Gamma = \frac{k_{-2}^{\frac{1}{2}}\, p^{\frac{1}{2}}}{k_2^{\frac{1}{2}} k' + k_{-2}^{\frac{1}{2}}\, p^{\frac{1}{2}}} \qquad 5\text{-}25$$

$$1 - \Gamma = \frac{k_2^{\frac{1}{2}} k'}{k_2^{\frac{1}{2}} k' + k_{-2}^{\frac{1}{2}}\, p^{\frac{1}{2}}} \qquad 5\text{-}26$$

THE HYDROGEN ELECTRODE

We have two cases:

Case 1:
$$k_2^{1/2} k' \gg k_{-2}^{1/2} p_{H_2}^{1/2} \qquad 5\text{-}27$$

From Eq. 5-25, $\Gamma \to 0$, and from Eq. 5-26, $1 - \Gamma \cong 1$. Therefore, Eq. 5-22 may be expressed:

$$i_1 = F k_1 c_{H^+} e^{-\beta \eta F / RT} \qquad 5\text{-}28$$

Case 2:
$$k_2^{1/2} \ll k_{-2}^{1/2} p_{H_2}^{1/2} \qquad 5\text{-}29$$

$\Gamma \cong 1$ in Eq. 5-25, and from Eq. 5-26:

$$1 - \Gamma \cong \frac{k_2^{1/2} k'}{k_{-2}^{1/2} p^{1/2}} \qquad 5\text{-}30$$

From Eqs. 5-30 and 5-22:

$$i_1 = F k_1 c_{H^+} k' \frac{k_2^{1/2}}{k_{-2}^{1/2} p^{1/2}} e^{-\beta \eta F / RT} \qquad 5\text{-}31$$

Here Γ did not introduce any additional potential dependence.

FAST DISCHARGE - SLOW RECOMBINATION, LANGMUIR

For the rate:

$$i_2 = k_2 (k'\Gamma)^2 \qquad 5\text{-}32$$

Assuming the discharge step is in equilibrium:

$$k_1 c_{H^+} (1 - \Gamma) e^{-\beta \eta F / RT} = k_{-1} k' \Gamma e^{+(1-\beta) \eta F / RT} \qquad 5\text{-}33$$

and:

$$\frac{\Gamma}{1 - \Gamma} = \frac{k_1 c_{H^+} e^{-\eta F / RT}}{k_{-1} k'} \qquad 5\text{-}34$$

or:

$$\Gamma = \frac{k_1 c_{H^+} e^{-\eta F/RT}}{k_1 c_{H^+} e^{-\eta F/RT} + k_{-1} k'} \qquad 5\text{-}35$$

Again, there are two cases:

Case 1, at low potentials:

$$k_1 c_{H^+} e^{-\eta F/RT} \ll k_{-1} k' \qquad 5\text{-}36$$

from Eq. 5-35:

$$\Gamma \cong \frac{k_2 c_{H^+} e^{-\eta F/RT}}{k_{-1} k'} \qquad 5\text{-}37$$

Substituting Eq. 5-37 into Eq. 5-32, we get:

$$i_2 = F k_2 \frac{k_1^2}{k_{-1}^2} c_{H^+}^2 e^{-2\eta F/RT} \qquad 5\text{-}38$$

Case 2, at high potentials:

$$k_1 c_{H^+} e^{-\eta F/RT} \gg k_{-1} k' \qquad 5\text{-}39$$

From Eq. 5-35, $\Gamma \cong 1$, therefore:

$$i_2 = F k_2 (k')^2 \qquad 5\text{-}40$$

here at low potentials, or low Γ, there is a potential dependence in the rate expression.

SLOW DISCHARGE - FAST ELECTROCHEMICAL, LANGMUIR

The rate is given:

$$i_1 = 2F \, c_{H^+} (1 - \Gamma) \, e^{-\beta \eta F/RT} \qquad 5\text{-}41$$

Taking the electrochemical step to be in equilibrium:

$$k_3 k' \Gamma \, c_{H^+} e^{-\beta \eta F/RT} = k_{-3} p \, (1 - \Gamma) \, e^{(1-\beta)\eta F/RT} \qquad 5\text{-}42$$

$$\frac{\Gamma}{1 - \Gamma} = \frac{k_{-3} p}{k_3 k' \, c_{H^+} e^{-\eta F/RT}} \qquad 5\text{-}43$$

THE HYDROGEN ELECTRODE

$$\Gamma = \frac{k_{-3} p}{k_3 k' c_{H^+} e^{-\eta F/RT} + k_{-3} p} \qquad 5\text{-}44$$

$$1 - \Gamma = \frac{k_3 k' c_{H^+} e^{-\eta F/RT}}{k_3 k' c_{H^+} e^{-\eta F/RT} + k_{-3} p} \qquad 5\text{-}45$$

Once again, we have two cases.

Case 1, at fairly high overpotentials:

$$k_3 k' c_{H^+} e^{-\eta F/RT} \gg k_{-3} p \qquad 5\text{-}46$$

From Eq. 5-44, $\Gamma \to 0$, and from Eq. 5-45, $(1 - \Gamma) \cong 1$, and we have for the rate:

$$i_1 = 2F\, k_1\, c_{H^+}\, e^{-\beta \eta F/RT} \qquad 5\text{-}47$$

Case 2, low potentials, if:

$$k_3 k' c_{H^+} e^{-\eta F/RT} \ll k_{-3} p \qquad 5\text{-}48$$

in Eq. 5-44, $\Gamma \cong 1$, and 5-45:

$$1 - \Gamma \cong \frac{k_3 k' c_{H^+}}{k_{-3} p} e^{-\mu F/RT} \qquad 5\text{-}49$$

putting Eq. 5-49 into 5-41:

$$i_1 = 2F k_1 c_{H^+} 2 \cdot \frac{k_3 k'}{k_{-3} p} e^{-(1 + \beta)\eta F/RT} \qquad 5\text{-}50$$

We now move on to some coupled mechanisms, where the reaction is fast enough that equilibrium does not apply. We use steady state rather than equilibrium conditions to obtain the quantity, Γ.

COUPLED DISCHARGE - ELECTROCHEMICAL, LANGMUIR

At steady state, the rates at the discharge and electrochemical steps are equal:

$$k_1 c_{H^+} (1 - \Gamma) e^{-\beta \eta F/RT} = k_3 c_{H^+} k' \Gamma e^{-\beta \eta F/RT} \qquad 5\text{-}51$$

and:

$$\frac{\Gamma}{1 - \Gamma} = \frac{k_1}{k_3 k'} \qquad 5\text{-}52$$

$$\Gamma = \frac{k_1}{k_1 + k_3 k'} \qquad 5\text{-}53$$

$$1 - \Gamma = \frac{k_3 k'}{k_1 + k_3 k'} \qquad 5\text{-}54$$

Case 1, if $k_1 \ll k_3 k'$, then from Eq. 5-53, $\Gamma \cong k_1/k_3 k' \to 0$, and from Eq. 5-54, $(1 - \Gamma) \cong 1$:

$$i_1 = 2F k_1 c_{H^+} e^{-\beta \eta F/RT} \qquad 5\text{-}55$$

Here we say that discharge is rate determining because Eq. 5-55 contains k_1, the rate constant for the discharge step. Case 2, if $k_1 \gg k_3 k'$, we have from Eq. 5-52, $\Gamma \cong 1$, and from Eq. 5-54:

$$1 - \Gamma \cong \frac{k_3 k'}{k_1} \qquad 5\text{-}56$$

putting Eq. 5-56 into Eq. 5-41:

$$i_3 = F k_3 k' c_{H^+} e^{-\beta \eta F/RT} \qquad 5\text{-}57$$

Here, the electrochemical step is rate determining.

SLOW DISCHARGE - FAST ELECTROCHEMICAL, TEMPKIN

We have for the rate:

$$i_1 = 2F k_1 c_{H^+} e^{-\beta r \Gamma/RT} e^{-\beta \eta F/RT} \qquad 5\text{-}58$$

Here we are including $(1 - \Gamma)$ in k_1 because these changes will be small compared to the changes in $e^{-\beta r \Gamma/RT}$. Taking the electrochemical step to be in equilibrium:

$$c_{H^+} k_3 e^{+(1-\beta) r \Gamma/RT} e^{-\beta \eta F/RT} = p\, k_{-3} e^{-\beta r \Gamma/RT} e^{(1-\beta) \eta F/RT} \qquad 5\text{-}59$$

$$e^{r \Gamma/RT} = \frac{k_{-3}}{k_3} \cdot \frac{p}{c_{H^+}} e^{\eta F/RT} \qquad 5\text{-}60$$

THE HYDROGEN ELECTRODE 71

$$e^{-\beta r\Gamma/RT} = (\frac{k_3}{k_{-3}} \cdot \frac{c_{H^+}}{p})^\beta e^{-\beta\eta F/RT} \qquad 5\text{-}61$$

Putting Eq. 5-61 into Eq. 5-58:

$$i_1 = 2Fk_1 (\frac{k_3}{k_{-3}p})^\beta c_{H^+}^{(1-\beta)} e^{-2\beta\eta F/RT} \qquad 5\text{-}62$$

The potential dependence and reaction order with respect to H^+ are both different from those obtained with Langmuir conditions.

FAST DISCHARGE - SLOW ELECTROCHEMICAL, TEMPKIN

For nonactivated desorption, we may write:

$$i_1 = 2Fk_3 c_{H^+} e^{r\Gamma/RT} e^{\beta\eta F/RT} \qquad 5\text{-}63$$

with the discharge step in equilibrium:

$$k_1 c_{H^+} e^{-\beta r\Gamma/RT} e^{-\beta\eta F/RT} = k_{-1} e^{(1-\beta)r\Gamma/RT} \cdot e^{(1-\beta)\eta F/RT} \qquad 5\text{-}64$$

$$e^{r\Gamma/RT} = \frac{k_1}{k_{-1}} c_{H^+} e^{-\eta F/RT} \qquad 5\text{-}65$$

$$e^{(1-\beta)r\Gamma/RT} = (\frac{k_1}{k_{-1}} c_{H^+})^{1-\beta} e^{-(1-\beta)VF/RT} \qquad 5\text{-}66$$

and putting Eq. 5-65 into Eq. 5-63:

$$i_3 = 2F \frac{k_3 k_1}{k_{-1}} c_{H^+}^2 e^{-(1+\beta)\eta F/RT} \qquad 5\text{-}67$$

For activated desorption, i.e., a desorption step involving a free energy of activation in addition to the free energy for the desorption process:

$$i_3 = 2Fk_3 c_{H^+} e^{(1-\beta)r\Gamma/RT} e^{-\beta\eta F/RT} \qquad 5\text{-}68$$

From Eq. 5-66 and 5-68:

$$i_3 = 2Fk_3 c_{H^+}^{2-\beta} (\frac{k_1}{k_{-1}})^{1-\beta} e^{\eta F/RT} \qquad 5\text{-}69$$

COUPLED DISCHARGE - ELECTROCHEMICAL, TEMPKIN

For nonactivated adsorption, using the steady state assumption:

$$k_1 c_{H^+} e^{-\beta r\Gamma/RT} e^{-\beta\eta F/RT} = k_3 c_{H^+} e^{r\Gamma/RT} e^{-\beta\eta F/RT} \qquad 5\text{-}70$$

$$e^{-(1+\beta)r\Gamma/RT} = \frac{k_3}{k_1} \qquad 5\text{-}71$$

$$e^{-\beta r\Gamma/RT} = \left(\frac{k_3}{k_1}\right)^{\beta/(1+\beta)} \qquad 5\text{-}72$$

$$i = 2Fk_1 \left(\frac{k_3}{k_1}\right)^{\beta/(1+\beta)} c_{H^+} e^{-\beta\eta F/RT} \qquad 5\text{-}73$$

For activated adsorption:

$$k_1 c_{H^+} e^{-\beta r\Gamma/RT} e^{-\beta\eta F/RT} = k_3 c_{H^+} e^{(1-\beta)r\Gamma/RT} e^{-\beta\eta F/RT} \qquad 5\text{-}74$$

$$e^{-r\Gamma/RT} = \frac{k_3}{k_1} \qquad 5\text{-}75$$

$$e^{-\beta r\Gamma/RT} = \left(\frac{k_3}{k_1}\right)^{\beta} \qquad 5\text{-}76$$

and from Eqs. 5-76 and 5-58:

$$i = 2F \, (k_1)^{(1-\beta)} \, k_3^{\beta} c_{H^+} e^{-\beta\eta F/RT} \qquad 5\text{-}77$$

Here we have mixed control, i.e., there is no rate determining step.

SLOW DISCHARGE - FAST RECOMBINATION, TEMPKIN

Since coverage is dependent only on pressure of hydrogen gas, we have the same kinetics as with the Langmuir case. Table 5.2 lists the results of the mathematics.

Much work has been done to evaluate the HER mechanism on various metals. Table 5.3 lists the probable mechanisms for several metals.

THE HYDROGEN ELECTRODE 73

Table 5.2. Tafel Slopes for Various Reaction Mechanisms

$$-\partial \frac{\partial \eta}{\log i}$$

	Langmuir Non-Act.	Langmuir Act.	Tempkin
Slow Discharge - Fast Recombination	$\frac{2RT}{F}$	$\frac{2RT}{F}$	$\frac{2RT}{F}$
Slow Discharge - Fast Electrochemical	$\frac{2RT}{F}$	$\frac{RT}{F}$	$\frac{RT}{F}$
Fast Discharge - Slow Recombination	$\frac{RT}{2F}$	$\frac{RT}{2F}$	$\frac{RT}{F}$
Fast Discharge - Slow Electrochemical	$\frac{2RT}{3F}$	$\frac{2RT}{3F}$	$\frac{RT}{F}$
Coupled Discharge - Recombination	$\frac{2RT}{F}$	$\frac{5RT}{2F}$	$\frac{3RT}{F}$
Coupled Discharge - Electrochemical	$\frac{2RT}{F}$	$\frac{2RT}{F}$	$\frac{2RT}{F}$

Source: J. McBreen and M.A. Genshaw, <u>Stress Corrosion Cracking</u>, Proceedings of a conference at Ohio State University, Columbus, Ohio, September 1967. Reprinted by permission of the National Association of Corrosion Engineers.

Taking the absolute coverages from the last bi-electrode experiment of chapter 2, we can do an interesting calculation for the hydrogen electrode rate constants in the instance of a palladium substrate. From table 5.3, palladium metal exhibits fast discharge - slow recombination. With Eq. 5-32, we get:

$$k_2 k'^2 = i_2/\Gamma^2 \qquad \qquad 5-78$$

where (from table 2.1 and Eq. 2-62) $i_2 = 1.1 \mu A$ and $\Gamma = 1.3 \times 10^{-4}$ coul cm^{-2}. Also, $\eta = +212$ mV. Thus:

Table 5.3. Reaction Mechanisms for Various Metals

						Al		
						D		
Ti		Mn	Fe	Ni	Cu	Ga		
D		A	E, C	A	A or D	A		
	Nb	Mo	Rh	Pd	Ag	Cd	Sn	
	D	D	B	B	C or D	A	A	
	Ta	W	Ir	Pt	Au	Hg	Tl	Pb	Bi
	D	A or D| B	B	D	A	A	A	D

A — Slow Discharge, Fast Recombination
B — Fast Discharge, Slow Recombination
C — Slow Discharge, Fast Electrochemical
D — Fast Discharge, Slow Electrochemical
E — Coupled Discharge Recombination

Source: J. McBree and M.A. Genshaw, *Stress Corrosion Cracking*, Proceedings of a conference at Ohio State University, Columbus, Ohio, September 1967. Reprinted by permission of the National Association of Corrosion Engineers.

$$k_2 k'^2 = 65 \text{ coul}^{-1} \text{ cm}^2 \text{ sec}^{-1} \qquad 5\text{-}79$$

Also, from Eq. 5-37, we get:

$$k_1/k_{-1} k' = \Gamma/c_{H^+} \, e^{-VF/RT} \qquad 5\text{-}80$$

where $\Gamma = 1.3 \times 10^{-4}$ coul cm^{-2}, $c_{H^+} = 10^{-4}$ moles cm^{-3}, and $\eta = +212$ mV. Thus:

$$k_1/k_{-1} k' = 0.77 \text{ coul cm mole}^{-1} \qquad 5\text{-}81$$

All this is at room temperature.

These calculations on the rate constants for hydrogen evolution would not have been possible without the absolute coverages of chapter 2 (table 2.1).

THE HYDROGEN ELECTRODE

Such calculations for other metals and mechanisms would be of interest; however, it is not presently known for which situations the experiments would be possible. For example, as previously discussed, these experiments would not be possible with the hydrogen-iron system.

STATISTICS OF THE REVERSIBLE HYDROGEN ELECTRODE AND IRREVERSIBLE EFFECTS

From thermodynamics of the reversible hydrogen electrode come two concepts: (1) the reversible potential depends on two quantities - hydrogen gas pressure and hydrogen ion activity in the electrolyte - only, and (2) the electrode material plays no role in what happens at the electrode except by the action of its surface (Bockris & Reddy, 1970), i.e., bulk properties should play no part.

Contrary to this, it will be shown that, in circumstances of high concentration and energy, hydrogen dissolved in the bulk of the electrode material can contribute to irreversible behavior of the hydrogen electrode.

Starting with Eq. 4-24 for ϕ and evaluating the constants for metal-hydrogen system, assuming a full monolayer of hydrogen, we obtain:

$$\phi = (3 \times 10^{-9})\ RT - a_H\ 2.5 \times 10^{-9}\ \exp\ (w_H/kT)\ RT \qquad 5\text{-}82$$

where units for ϕ might be ergs cm^{-2} and a_H is moles of H cm^{-3}. In Eq. 5-82, the first term is for a surface monolayer of H atoms, and the second for the dilute gas of protons in the bulk electrode material.

Taking the case of platinum, we approximate the w_H term as the heat of solution. Actually, w_H is the protonic work function for the process:

H (infinite dispersion) = H (dissolved in the metal)   5-83

while heats of solution are for the process:

½ H_2 (gas) = H (dissolved in the metal)   5-84

the situation of external hydrogen. Here, take as an approximation that Eqs. 5-83 and 5-84 are the same (Fullenwider, 1974).

There are two reported heats of solution for hydrogen in platinum: 8.9 Kcal and 24.3 Kcal (Richardson, Nicol, & Parnell, 1904; Gileadi, Fullenwider, & Bockris, 1966). The bulk solubility of hydrogen at the reversible potential is 10^{-5} (mole H cm^{-3}) (Gileadi et al., 1966).

Substituting the value of the solubility at the reversible potential and considering the heats of solution, it can be seen that the bulk term is about equal to the surface term for the first heat of solution, and that the bulk term completely predominates with the second value. In these two situations, then, thermodynamics is contradicted, and we will have irreversible behavior. However, the value of a_H cited for platinum was in the presence of As_2O_3, a hydrogen permeation promoter, and in the absence of this, the solubility would be much lower. Therefore, the irreversible behavior observed in the presence of many cations and anions is probably due to their hydrogen permeation promoting properties.

From these considerations, it is also possible to derive isotherms relating surface coverage to bulk hydrogen concentration in the metal. Isotherms usually relate surface coverage to electrochemical potential, generally expressed as a function of concentration in a bulk electrolyte solution or potential across the interface (Gileadi, 1967).

Here we have (Gileadi, 1967):

$$RT\Gamma_H = \partial\phi/\partial\ln a_H \qquad 5\text{-}85$$

THE HYDROGEN ELECTRODE

where Γ_H is surface coverage with hydrogen, and the rest of the symbols have the same meaning as before. The quantity a_H is a function of the quantities mentioned above for isotherms, but here their functionality will be hidden.

Rewriting Eq. 4-24 as:

$$\phi = K \Gamma_H + K' a_H \exp(w_H/kT) \qquad 5\text{-}86$$

we get from Eq. 5-85:

$$RT \Gamma'_H = K a_H \partial \Gamma'_H/\partial a_H + K' a_H \exp(w_H/kT) \qquad 5\text{-}87$$

which is a linear, first order differential equation with solution:

$$\Gamma'_H = C a^{2/3} - \frac{3}{2} k' a_H \exp(w_H/kT) \qquad 5\text{-}88$$

where C is a constant of integration. Eq. 5-88 relates surface concentration to a dilute gas of protons dissolved in the metal for the situation of equilibrium and no interactions between the protons in the gas.

Similarly, Eq. 4-26 may be written:

$$\phi = K \Gamma'_H + K' a_H' \exp[w_H + N_H w_{HH}/\alpha N)/kT] \qquad 5\text{-}89$$

where:

$$a_H' = N_H/(\alpha N - N_H) V \qquad 5\text{-}90$$

Again, from Eq. 5-83:

$$\Gamma'_H/a_H'^{2/3} = C - \int K'/a'^{2/3} K \exp[(w_H + N_H w_{HH}/\alpha N)/kT]$$
$$[1 + a_H' w_{HH} V/kT (1 + V a_H')^2] da_H' \qquad 5\text{-}91$$

Eq. 5-91 extends Eq. 5-88 into regions where interactions between the absorbed protons in the metal are not negligible. Equilibrium is still assumed between absorbed and adsorbed protons.

6
Catalysis

INTRODUCTION

A discussion of catalysis could be a lengthy matter if it were not for the fact that most are company proprietory. Many times, a company in possession of a new, useful catalyst will not even take out a patent on it because they do not want other companies to know what they are doing.

The catalysts discussed here are of a type in which the present author is interested because they relate to the energetics and permeation of metals by hydrogen and oxygen. Two types of catalysts will be discussed: (1) hydrogen oxidizing catalysts, and (2) oxygen reducing catalysts. The low temperature hydrogen-oxygen fuel cell is such an integral part of the hydrogen economy that it is all right to discuss oxygen reducing catalysts here. Oxygen also permeates metals (Schuldiner & Warner, 1965). Aside from numerous other applications for such catalysts, these catalytic systems are used in the two electrodes of the low temperature hydrogen-oxygen fuel cell. Figure 6.1 is a schematic of such a fuel cell. A 4.8 megawatt fuel cell installation is currently being constructed in New York City by United Technologies Power Systems Division. For an order of magnitude estimate of

CATALYSIS   79

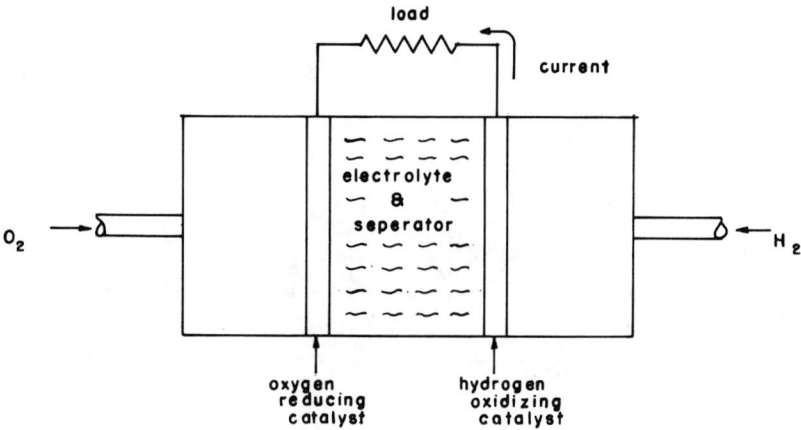

Fig. 6.1. Schematic of a Hydrogen Oxygen Fuel Cell

just how much power this is, it may be taken that 200 average homes may be supplied with electric power per megawatt. See figure 6.2 for an artist's rendering of this site. This type of fuel cell uses oxygen from the atmosphere and cracks oil or natural gas to obtain hydrogen. It operates at about 180°C, and the electrolyte is phosphoric acid. Platinum, an expensive metal, is used as catalyst on both sides. It would be a step forward to find a less expensive substitute for this catalyst. Actually, it has been found that ruthenium is a better catalyst for the oxygen reduction, but less noble than platinum. The oxygen side is by far the slower of the two, and it is here that better catalysts are needed the most.

Another factor is the cost of hydrogen. Currently, hydrogen is being produced commercially by the thermal decomposition of hydrocarbons, as in the New York City fuel cell site. Other methods, such as, for example, electrolysis of water, are too expensive at present to compete with the hydrocarbon approach. This problem is envisioned to find a solution when "extra energy" becomes available, e.g., solar

Fig. 6.2. Artist's Rendering of the New York City Fuel Cell Site

energy for the electrolysis of water which is being worked on by many researchers; hydrogen fusion reactors are another possibility for low cost energy, thought to be available around the year 2000. Besides these two approaches are the usually taked about windmills, remotely placed fission reactors, geothermal energy, and so on.

HYDROGEN OXIDIZING CATALYSTS

In a fuel cell, at the hydrogen oxidizing side, we have an example of external hydrogen, discussed in chapter 4, i.e., hydrogen gas is bubbled in around the solid catalyst – solution interface. The final equation for this situation was, in the dilute case:

$$\frac{N_H}{V_H} = \left(\frac{m_H}{16\pi A^2 h^2 kT}\right)^{1/4} p^{1/2} e^{-(w_H + \frac{1}{2}\chi_d)/kT} \qquad 6\text{-}1$$

and in the concentrated case:

$$\frac{\theta''}{1-\theta''} = \left(\frac{p}{kT}\right)^{1/2} e^{-(w_H + 2\theta'' w_{HH} + \frac{1}{2}\chi_d)/kT}$$

$$\times \left\{\frac{(2\pi 2 m_H kT)^{3/2}}{h^3} \frac{8\pi^2 A kT}{2 h^2}\right\}^{-1/2} \qquad 6\text{-}2$$

There is some evidence that the concentration of dissolved hydrogen just inside the surface, c_o, of metal hydrogen systems is greater than previously thought.

First, figure 2.7 illustrates qualitatively the concentration profile for the palladium-hydrogen system, resulting from the experiment of chapter 2. From this figure, it can easily be seen that c_o might have been underestimated in the interpretation of experiments where it was thought that the concentration profile was linear. In fact, the results of such experiments would come from the estimation of the slope of lines which were tangents to the curve in figure 2.7 at $x = \delta$. It is clear that this would result in large underestimations in the calculated value of c_o.

Also, from the result of chapter 4 for the spreading pressure of internal hydrogen surfaces:

$$\phi = (3 \times 10^{-9}) RT - a_H [2.5 \times 10^{-9} \exp(w_H/kT)] RT \qquad 6\text{-}3$$

it can be seen that the bulk (second in 6-3) term can make large contributions to the spreading pressure at or near the surface of the platinum-hydrogen system, if circumstances are right. This would cause greater c_o's.

Finally, the pumping action of rate limitation from chapter 3 can be imagined to result in large c_o's.

As we stated in chapter 4, Eq. 6-1 here, the quantity $w_H + \chi_d$ may be approximated as the heat of solution for hydrogen in the metal. It should be remembered that w_H can be plus or minus as the heat of solution is endothermic and exothermic, respectively, i.e., w_H should have the same sign as the heat of solution, χ_d is always positive.

All metals have some catalytic properties because hydrogen enters almost all metals, causing embrittlement. Table 6.1 lists some heats of solution.

Table 6.1. Heats of Solution of Hydrogen in Various Metals

Metal	ΔH, kcal (mole H)$^{-1}$	Reference
Ni	3.60	(Ebisuzaki, Kass, & O'Keefe, 1967)
Pd	-9.5	(Maeland & Gibb, 1961)
Pt	8.9	(Richardson, Nicoll, & Parnell, 1904)
Pt	24.3	(Ham, 1932)
Fe	7.04	(Bryan & Dodge, 1963)
Fe	5.0	(Bockris, Genshaw, & Fullenwider, 1972)
Cu	9.0	(Flanagan & Oates, 1972)
Y	-20	(Flanagan & Oates, 1972)
Sc	-20	(Flanagan & Oates, 1972)

CATALYSIS

Table 6.2 lists some values of the concentration of hydrogen dissolved in the bulk phase, a_H, for several metals at the reversible potential which corresponds to the solubility of H_2 at one atmosphere.

Table 6.2. Solubilities of Hydrogen in
Various Metals at the Reversible Potential

Metal	a_H (mole H cm^{-3})	Reference
Pt	1.0×10^{-15}	(McBreen, Nanis & Beck, 1966)
Ni	1.4×10^{-5}	(Bockris et al., 1972)
Pd	2.7×10^{-4}	(Bockris et al., 1972)
Fe	7.0×10^{-9}	(Bockris et al., 1972)

From tables 6.1 and 6.2, and Eq. 4.3, it can be seen that it is only metals with w_H far in the endothermic region (Pt), that the bulk term becomes large.

PLATINUM AND THE SEARCH FOR SUBSTITUTE CATALYSTS

Platinum has long baffled researchers. One of the properties of platinum is that it is very inert, and just this property has been the object of many investigations trying other systems which are simply inert, in, for example, the 180° C phosphoric acid of the low temperature hydrogen-oxygen fuel cell.

Another property of platinum is the splitting of the hydrogen molecule bond by the surface. This is the property which will be of more interest here, i.e., it will be assumed that the reaction environment can be varied to suit a less noble but more active catalyst.

The approach here will be very simple. the two values of ΔH for platinum in table 6.1 is the first hint of our approach. It will be proposed here that the H_2 bond splitting property of Pt is due to the presence of closely adjacent sites of widely differing heats of solution for hydrogen at the surface or just beneath. To simulate this, two metal systems will be chosen that are as dissimilar energetically (with respect to heats of solution) as possible, acting through the pumping mechanism of rate limitation to expose the hydrogen molecule bond to the two metals, breaking it.

From table 6.1, we choose two systems: platinum-yttrium and copper-yttrium; scandium is very expensive. As for the type of system, the epitaxial layer of copper on yttrium and the same of platinum on yttrium seemed most promising (Fullenwider, 1977). Alloys and composites are not promising because of leaching between the two metals. One investigator (Bocciarelli, 1967), for example, was led by his considerations to a gold-copper alloy for an oxygen reduction catalyst which, as was found out later, produced an electrochemical couple to give currents indicating good catalytic activity but, in time, resulted as a Raney gold due to the dissolution of the copper. Composites suffer from the same problem.

The epitaxial layer can be made in powder form by the use of electroless plating solution, available for copper and platinum. Figure 6.3 illustrates a particle of such a powder. Yttrium can be ground to a fine powder by saturating it with embrittling hydrogen. It smears, while grinding, in the absence of hydrogen.

With this catalyst, hydrogen exposure is provided to the two metals by means of the rate limitation pumping action, while only one metal is exposed to the reaction environment. Thus, there will be no leaching.

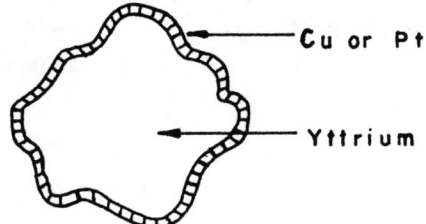

Fig. 6.3. Particle of Epitaxial Catalyst Powder

OXYGEN REDUCING CATALYSTS

Oxygen, as hydrogen, penetrates the surface of platinum electrodes (Schuldiner & Warner, 1965). Oxygen-platinum has not been studied extensively as a diffused system as hydrogen has, though. The oxygen molecule has a double bond, of strength 118.86 kcal mole^{-1}, and it is difficult to break, adding to the irreversibility of its reaction in fuel cells. Another difficulty with the oxygen side is that very few metals are stable enough to withstand the environment of the oxygen reduction electrode. Platinum is considered a good catalyst for this reaction only because it does not dissolve. Actually, the exchange current density for the reaction at platinum of 10^{-10} amp cm^{-2} indicates that platinum is really a very poor catalyst. It is proposed here that better catalysts are possible with the epitaxial growth approach, the same approach as used with the treatment of hydrogen oxidizing catalysts.

Our approach, neglecting the hydrogen-peroxide problem, will be to look for an activated ruthenium or platinum, that is, look for substrated on which to put an epitaxial layer of these metals.

There are no heats of solution available in this case. What we will use here is bond strengths of metal oxides, looking for large energetic differences again between ruthenium and platinum and (of practical cases) the substrate metal, resulting in a situation where the oxygen molecule bond will be broken by the catalyst.

Metals of suitable characteristics, besides the first three in table 6.3, lie between $\sim$ 170 kcal mole^{-1} and $\sim$ 70 kcal mole^{-1}.

Table 6.3. Bond Strengths of some Metal Oxides

System	Bond Strength (kcal mole^{-1})
O-Th	192 ± 10
O-Nb	189 ± 10
O-Zr	181 ± 10
O-Ru	115 ± 15
O-Pt	83 ± 8

Source: *Handbook of Chemistry and Physics* (55th Ed.) Chemical Rubber Co., 1974.

In table 6.3 are listed some metals for substrates, ruthenium and platinum. Thorium has the greatest bond strength but is radioactive. Niobium and zirconium are the choices for substrates. These combinations are patented (Boehm, Treptow, Wunsch, Kiener, Meger, & Csizi, 1979; Heikel & Leddy, 1980; Spaziante & Nidola, 1980) in the sheet form from chlor-alkali work. Platinum on niobium and zirconium substrates looks good, also, and might stand up better in acid and high temperature environments.

At present, there is no electroless plating solution available for ruthenium. It is still possible to make epitaxial powders, however, with the use of a rotating drum where electrical contact with a substrate powder is made through a cathode immersed in the substrate powder. There is, however, electroless platinum available (Leaman, 1972).

7
Storage and Purification of Hydrogen as Metal Hydrides

INTRODUCTION

As fossil fuels become scarce, hydrogen is believed by some to be the answer to the energy shortage which will result. For utility applications, such as fuel cells, it will be possible to pipe hydrogen from remotely placed nuclear reactors specifically for hydrogen production, more cheaply than electricity.

In other applications, such as vehicle propulsion, existing containment technology will have to be expanded. The storage of hydrogen is presently accomplished with the gas cylinders, and, at low temperatures, methods which are hazardous and clumsy for everyday use. It is currently thought that ternary metal hydride alloys in the powder form will provide a solution to this problem.

Most of the material in this chapter comes from a recent review (Reilly, 1979). The following alloys are discussed: Mg_2NiH_x, Mg_2CuH_x, $FeTiH_x$, $TiCrH_x$, and $LaNi_5H_x$. This choice is dictated by cost and availability factors.

There are three rules which such ternary alloys obey (Reilly, 1977):

STORAGE AND PURIFICATION OF HYDROGEN 89

1. At least one of the components of the binary metal alloy should be capable of reacting reversibly with hydrogen to form a stable binary hydride.
2. If the metal atoms are mobile at the temperature of formation, the resulting configuration will be the most favorable one.
3. At lower temperatures, where the metals atoms are not mobile, hydride phases will form which are similar to the initial compound.

Rule 1 is an empirical one (Van Mal, 1976; Van Mal et al., 1975), while 2 and 3 are based on thermodynamic principles related to the free energy and speed of diffusion.

There are some trends which are worth mentioning. One of these is the rule of reversed stability (Van Mal et al., 1975). In this, the formation of the hydride from addition of hydrogen to an alloy has an enthalpy change, ΔH.

$$\Delta H_{AB_n H_{2m}} = \Delta H_{AH_m} + \Delta H_{BH_m} - \Delta H_{AB_n} \qquad 7\text{-}1$$

where A and B are metals, and A is a stable hydride former. The success of this relation has been limited, e.g., does not apply to titanium alloys (Johnson & Reilly, 1978; Yamanaka, Saito, & Someno, 1975).

The free energy of formation of homologous series of alloy hydrides in the plateau region of the isotherm (Fig. 8.3) is found to be more negative as the interstice becomes larger. This relation has met with some success (Wagner, 1971).

An interesting relation for pure metal hydrides is that the product of the solubility in moles cm^{-3}, and the diffusion coefficient for a metal in cm^2 sec^{-1} is usually in the range of 10^{-13}. These properties may be compromised by the formation of ternary alloys. Looking at figures 7.1, 7.2, and 7.3, the iron-hydrogen system can be seen to be the fastest

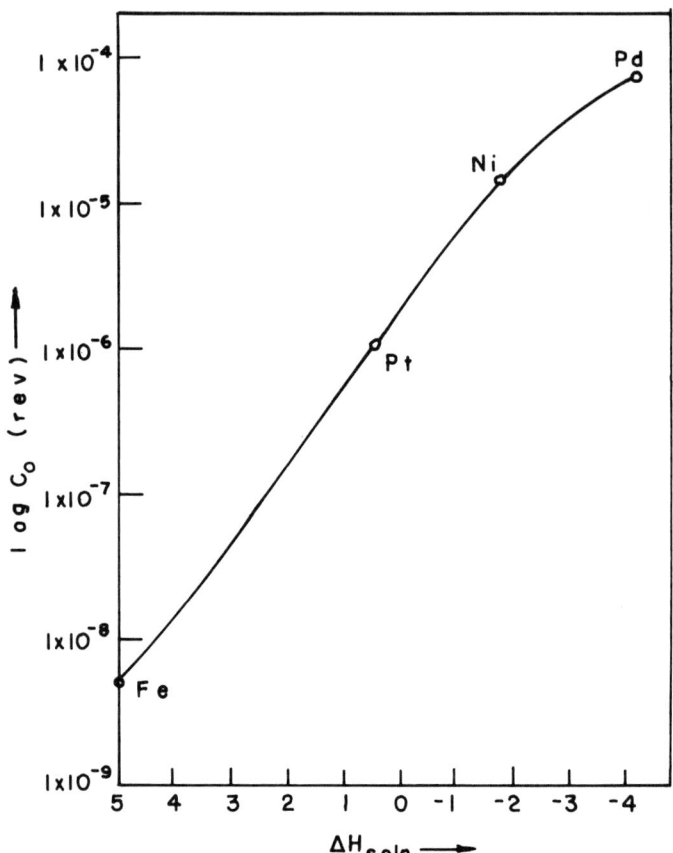

Fig. 7.1. Log (Solubility) at the Reversible Potential (Corresponding to the Reversible Potential) Versus Heat of Solution.

diffuser, and has the lower solubility, while it has been reported that the titanium has an extreme negative heat of solution (~ -10 kcal^2 mole^{-1}). Thus, this alloy combination of high diffusion coefficient and high solubility could be expected to have good properties as a hydrogen storer; the iron-nickel combination also appears promising. There are

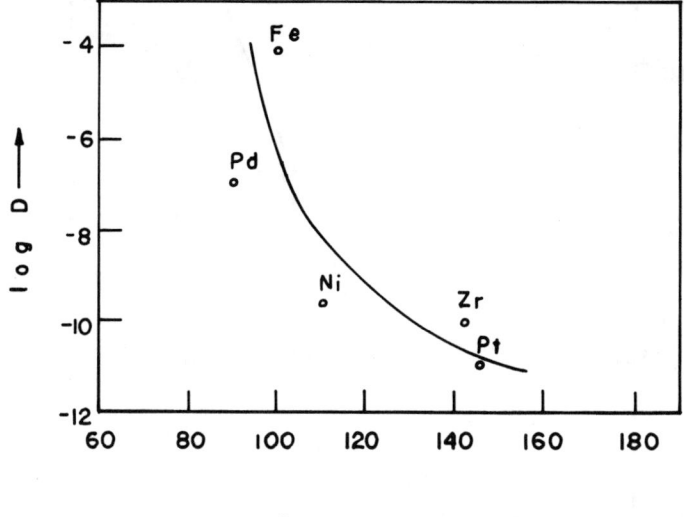

Fig. 7.2. Log (Diffusion Coefficient) Versus Latent Heat of Sublimation. Latent heat of sublimation is a measure of the energy required to stretch the metal-metal bond as hydrogen diffuses through the lattice.

also metals with even more negative heats of solution than titanium for hydrogen (Flanagan & Oates, 1972), but these, e.g., Y, Sc, etc. ($\Delta H \sim -20$ kcal mole), are rare and expensive.

The crystal structures of some alloys are of interest (Reilly, 1979).

At 298 Kelvins $FeTiD_{1.0}$ is orthorhombic, a = 2.956 Å, b = 4.543Å, and c = 4.388Å. At higher concentrations at the same temperature ($FeTiD_{1.9}$) of deuterium, the ternary alloy becomes monoclinic: a = c = 4.7044Å, b = 2.8301Å, and β = 96.57°.

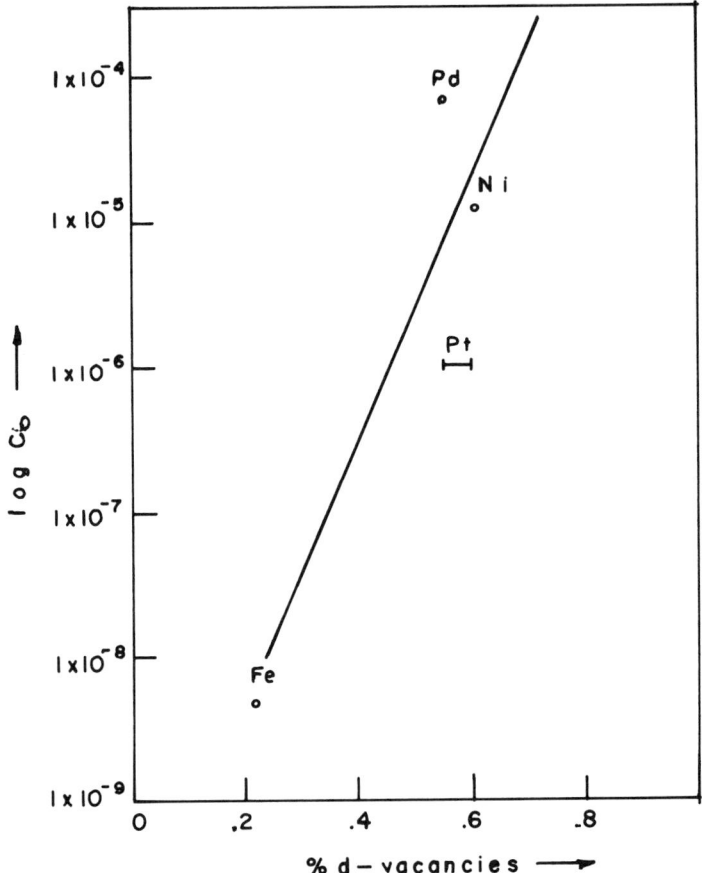

Fig. 7.3. Log (Solubility) at the Reversible Potential Versus % d-Vacancies. Solubility increases with % d-vacancies because of the greater room for the electron donated by the proton.

$LaNi_5H_6$ at 298 Kelvins is hexagonal with a = 5.440Å, and c = 4.310Å.

More data of this type would be useful.

STORAGE AND PURIFICATION OF HYDROGEN

IRON-TITANIUM ALLOYS

Of the alloy systems discussed (Reilly, 1979), the iron-titanium system shows the most promise, mainly from cost and availability factors. For discussion of the other alloys, Reilly's review can be consulted.

There are two alloys in the iron-titanium phase diagram: FeTi and Fe_2Ti (Reilly & Wiswall, 1974). Fe_2Ti is useless as a hydrogen storer because it dissolves hydrogen only at high temperatures and pressures.

FeTi forms two distinct hydrides according to the reactions:

$$2.13\ FeTiH_{0.1} + H_2 \rightleftarrows 2.13\ FeTiH_{1.04} \qquad 7\text{-}2$$

and:

$$2.20\ FeTiH_{1.04} + H_2 \rightleftarrows 2.20\ FeTiH_{1.95} \qquad 7\text{-}3$$

where initially the metals are saturated with hydrogen. The alloys are very brittle because of the action of hydrogen. Both iron and titanium embrittle, so this is no surprise. Exposure of the alloy to the atmosphere deactivates it, probably because of the formation of oxide films. Titanium forms an oxide which is very impermeable to hydrogen, while in the case of iron, the oxide film is fairly permeable to hydrogen. The alloy can be reactivated by charging with hydrogen. The reversibility of the hydrogen charging and outgassing cycle was tested 30,000 times with the temperature varying from 0° to 100° C (Reilly, 1978). There was no deterioration, the only change being in the magnetic susceptibility. The increase was thought to be due to the formation of iron clusters.

There seems to be another avenue of approach to these problems which has not yet been explored. It would seem that epitaxial powders of a type similar to the catalysts discussed in chapter 6 might be useful. Palladium, perhaps in a thin film, might help the hydrogen to get into the alloy faster and come out easier. There exists a palladium electroless plating solution, and a test of this effect would not be too difficult.

The alloys are also thought useful as purifiers of hydrogen from gas mixtures, but are easily poisoned, not only by oxygen in air, but by carbon monoxide. An epitaxial coating of some inert metal, such as platinum, might be useful here.

At present, most metal hydride studies are being examined on the basis of their energy storage capabilities. This work is still in its infancy.

8
Embrittlement

INTRODUCTION

The field of hydrogen embrittlement is a very active one with many theories (Flitt & Bockris, 1981). Heats of solution for hydrogen vary from 24 kcal mole^{-1} for platinum (Ham, 1932) to -20 kcal mole^{-1} for scandium and yttrium (Flanagan & Oates, 1972) while solubilities range from 10^{-7} moles cm^{-3} for iron to complete saturation of the available interstitial sites for hydrogen in the more exothermic occluding metals.

It seems unrealistic to expect a single model of hydrogen embrittlement to hold over such a wide range of variables. It is a certainty, though, that most metals, even over this wide spread of variables, do embrittle, sometimes catastrophically.

We begin again with the Barnacle Electrode where the work has become quantitative.

EMBRITTLEMENT STUDIES WITH THE BARNACLE ELECTRODE

"Indexing the degree of hydrogen embrittlement of 4340 steel using the Barnacle Electrode" has been accomplished (Deluccia & Berman, 1981). Figure 8.1 shows the resulting plot. This is a result which could contribute to other types of experi-

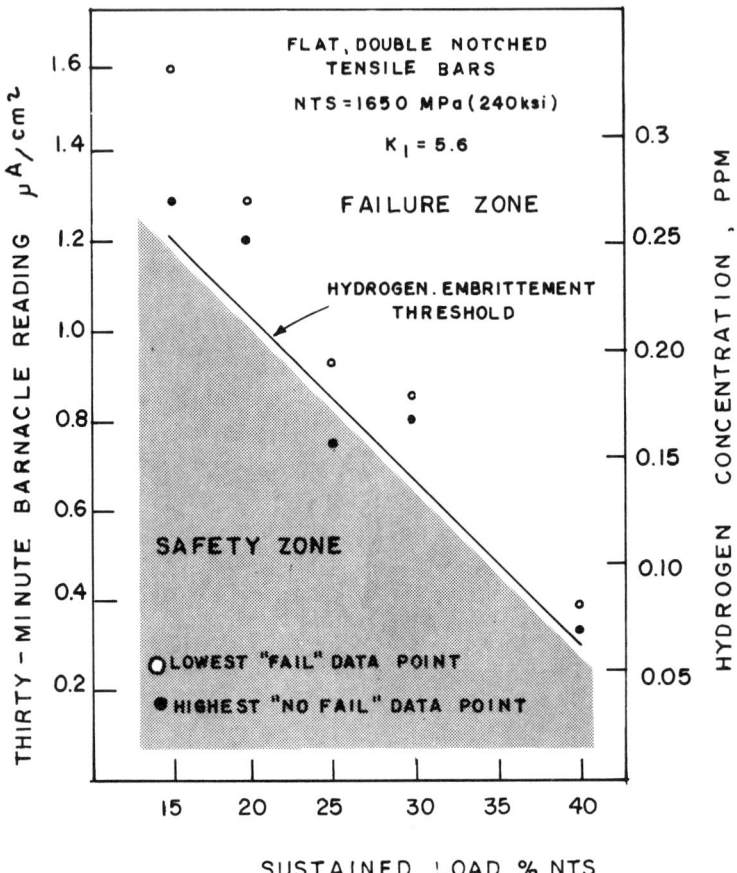

Fig. 8.1. Estimate of Threshold of Embrittlement as a Function of Concentration of Hydrogen and of Sustained Load (% NTS).

Source: D.A. Bermann, J.J. DeLuccia, and F. Mansfeld, "Barnacle Electrode: New Tool for Measuring Hydrogen in High Strength Steels," Metal Progress, May 1979. Reprinted by permission.

EMBRITTLEMENT   97

ments. DeLuccia and Berman's results are empirical in nature, and it would help matters if they could be related to a theory. In these experiments, the steel pieces were actually broken to determine where the onset of embrittlement occurred. Combined with the concept of "critical concentration" or "critical current density" (Beck, Bockris, McBreen, & Nanis, 1966), figure 8.2 could do a lot to clear the issue. Figure 8.2 shows qualitatively the concept of critical current density. The peaked shape is interpreted as being due to the breakup of the metal due to embrittlement, thus blocking the permeation current. We cannot do much here because the critical current density experiments were performed on armco iron, not 4340 steel. This area needs more work. In the meantime, the Barnacle Electrode is firmly established on a quantitative basis.

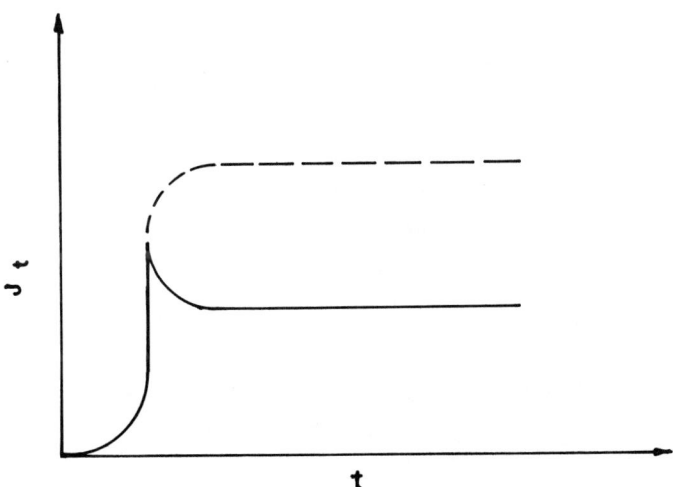

Fig. 8.2. Typical Transient Illustrating Critical Current Density

THEORIES OF EMBRITTLEMENT

Most of the beginning material for this section follows standard lines (Subramanyan, 1970) followed by a discussion of insights from statistical arguments.

There are three main arguments in the treatment of hydrogen embrittlement of iron and steel. The most popular is the pressure expansion approach. According to Benneck, Schenck, and Muller (1935), molecular hydrogen precipitates in internal voids or cracks in the lattice, developing extreme high pressures at which cracks spread causing premature failure. A different idea is put forth by Petch (1956). According to this, as hydrogen adsorbed on the internal surfaces of cracks and voids, the surface energy was lowered, causing embrittlement. According to a third approach by Troiano (1960), hydrogen embrittlement was caused by the accumulation of atomic hydrogen in regions of high triaxiality.

PRESSURE EXPANSION THEORIES OF EMBRITTLEMENT

First, we discuss Griffith's theory of brittle fracture (Griffith, 1929) which is based on thermodynamics. The equilibrium state of an elastic solid body, deformed by surface forces, is that in which the potential energy of the whole system is a minimum. If the system can pass from the unbroken to the broken situation involving a continuous decrease in potential energy, the equilibrium state, if equilibrium is possible, must be that in which the solid has been broken.

Formation of a crack in an elastic solid has to be done against the cohesive forces between the molecules or atoms on either side of the crack. The work appears as potential surface energy and, if the width of the crack is greater than the small distance referred to as "the radius of molecular action," the energy per unit area of the material is the surface energy.

If W is the strain energy released due to the formation of a crack of length 2l, and U the surface energy of the crack, then the total decrease of potential energy is (W-U), and the condition for spreading of the crack is:

$$\frac{\partial}{\partial l} (W-U) < 0 \qquad \qquad 8-1$$

where W is a function of the elastic constant of the solid, Poisson's ratio, dimensions of the crack, and the external stresses.

This relation has been used rather extensively to derive expressions for details about the propagation of cracks. The more recent theory of Beck, Bockris, McBreen, and Nanis (1966) is centered about the concept of the "critical current density," derived from experiments with the bi-electrode technique. In figure 8.2 is illustrated the shape of a diffusion transient, for a thin iron membrane, above the critical point. Above this critical current, transients are "peaked" and below, they have a normal, reproducible transient appearance as figure 2.2. This peaked shape is believed to be due to the onset of embrittlement. This effect occurs with identical characteristics in the case of polycrystalline armco iron, single crystal iron, and zone refined iron. In addition, the effect was independent of the electrolyte, and the formation of a chemical hydride was ruled out because x-ray investigations showed little or no change in lattice parameters upon saturation with hydrogen.

In more detail, the occurrence of the maximum in the permeation transient was explained as follows (Beck et al., 1966): when the concentration of hydrogen is low, it diffuses through the membrane without damaging it. Upon reaching the critical concentration in the metal, however, hydrogen molecules precipitate at some hiatus in the metal and cause it to expand to form cracks or blisters. This embrittling decreases the rate of permeation. According to Beck and his colleagues, the site for the nucleation of a crack is a pile-up of dislocations. The effect of grain boundary or impurity in the nucleation of cracks was ruled out because all three situations - polycrystal, zone refined, and single crystal iron - exhibited almost the same effect.

The propagation of a crack is according to the Griffith theory. It is approximated that the nucleus of the void is a Griffith crack of length 2l. Then the critical internal pressure, P_{cr}, of hydrogen for the propagation of the crack is:

$$P_{cr} = (\frac{2\gamma Y}{\pi l})^{\frac{1}{2}} \qquad \text{8-2}$$

where γ is the surface energy per unit area and Y the Young's modulus, and with Sievert's law:

$$c = K\, p^{\frac{1}{2}}\, e^{-\Delta H_s/RT} \qquad \text{8-3}$$

where p is the pressure of hydrogen in the crack, and ΔH_s is the heat of solution of hydrogen in the metal.

With Eqs. 8-2 and 8-3, it is possible to calculate the equivalent pressure corresponding to any concentration. If c_{cr} is the critical concentration of hydrogen corresponding to P_{cr}, then:

$$c_{cr} = (\frac{2\gamma Y}{\pi l})^{\frac{1}{2}} K\, e^{-\Delta H_s/RT} \qquad \text{8-4}$$

or:

$$c_{cr} = K'\, e^{-\Delta H_s/RT} \qquad \text{8-5}$$

In a limited temperature range, Y, γ, l, and K can be taken as constants. Therefore, c_{cr} for iron should increase with temperature, which is what is observed.

ΔH_s is calculated from the relation:

$$\Delta H_s = R \left(\frac{d \ln c_{cr}}{d\, 1/T}\right) \qquad 8\text{-}6$$

and was found to be 4 kcal per mole H. This value compares well with Sievert's value of 7 kcal. It is of interest that c_{cr} was found to decrease with increasing temperature in the case of platinum (Gileadi et al., 1966). This has not been explained.

THEORIES INVOLVING THE REDUCTION OF THE METAL COHESIVE STRENGTH

Petch (1956) presents another theory where a reduction of cohesive strength of a metal was taken to be due to adsorption, under Langmuir isotherm conditions. This theory is of interest for reasons to be discussed later in the chapter. It is not generally given much support by workers in the area (Flitt & Bockris, 1981).

Another proposed theory is that of Troiano (1960) in which embrittlement would be caused by the collection of hydrogen in regions of high stress, lowering the cohesive strength of the metal. This theory has been developed further by Oriani (1972).

More insight into surface effects will be obtained with the use of statistical models.

STATISTICAL THEORIES

It is not proposed here to argue against the obvious utility of the pressure expansion theories which are usually combined with the high equivalent pressure electrochemical mechanism for getting the hydrogen into the bulk of the metal. This approach to the problem is here to stay; but electrochemists, when asked, "What happens when there is no hydrogen overpotential?", usually try to envoke something like a thin layer of water at the surface of the metal, i.e., they cannot get by without this overpotential.

The pumping action of the constant flux boundary condition, as shown in chapter 3, partially does away with the need of overpotential, although it does not replace it.

So it is now possible to get hydrogen into the metal without electrochemical overpotentials, but when there are overpotentials, this type of effect will certainly predominate.

The approach to be taken in this section will be to regard the hydrogen as internal. Thus, we have from chapter 4, evaluating the constants for dilute metal-hydrogen systems:

$$\phi = (3 \times 10^{-9}) \, RT - (2.5 \times 10^{-9}) \, a_H \, RT \, \exp(w_H/kT) \qquad 8\text{-}7$$

and for more concentrated systems:

$$\phi = (3 \times 10^{-9}) \, RT - (2.5 \times 10^{-9}) \, (\frac{N_H}{\alpha N - N_H}) \, \exp$$

$$(w_H + N_H w_{HH}/\alpha N)/kT \qquad 8\text{-}8$$

In Eqs. 8-7 and 8-8, the first terms are surface contributions and the second bulk contributions. As the arguments of chapter 5 pointed out, concentrations at any surface will be greater than in the bulk. Here we will be dealing with internal surfaces, cracks, voids, and so on.

From Eq. 8-7, we can conclude that only for metals of extreme high endothermic heats will the bulk term make any contribution at all. Concentrations at cracks and voids, however, will be large. In addition to the arguments of chapter 2, we have the effect (Beck et al., 1966):

EMBRITTLEMENT

$$c_{H,\sigma_H} = c_{H,o}\, e^{\overline{V}_{H,\sigma_H}/RT} \qquad 2\text{-}30$$

where c_o is the stress-free concentration in the bulk of the metal, c_σ the concentration in regions of hydrostatic stress σ (σ will be greater in the vicinity of a crack or other internal surface), positive for tensile stress and negative for compressive stress, and $\overline{V}$ the partial molar volume of H.

This is illustrated in figure 8.3, a typical isotherm for a metal-hydrogen system at a surface. In the figure, the stress free bulk of the metal can be imagined to be close to c while at a crack or void with concentration, perhaps greater by an order of magnitude, the pressure will lie higher along C-D, the pressure of which increases very rapidly with concentration, embrittling the metal.

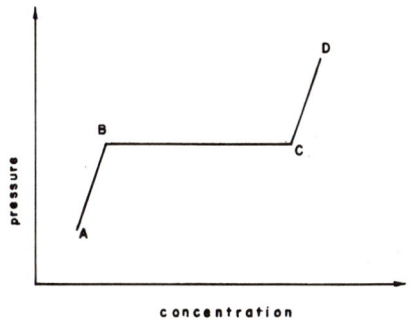

Fig. 8.3. Typical Metal Hydrogen Isotherm

Mathematically, this can follow from Eq. 8-8 for endothermic occluders. The factor from Eq. 8-8:

$$\frac{N_H}{\alpha N - N_H} \qquad 8\text{-}9$$

as long as the heat factor:

$$\exp\,(w_H + N_H w_{HH}/\alpha N)/kT \qquad 8\text{-}10$$

is not too exothermic, then as a semiclassical "band" (Gosar, 1964) fills up ($\alpha N \rightarrow N_H$) causing the factor 8-9 of Eq. 8-8 to have large values in the spreading pressure, ϕ, embrittling the metal. It should be noted that here we are dealing with a surface effect which could be interpreted as the decohesion effect (Oriani, 1972; Petch, 1956). This could play a part at lower values of ϕ, but gives way to pressure expansion at higher values of ϕ where equilibrium between internal hydrogen and molecular hydrogen within cracks and voids is set up. It would appear, however, that the decohesion will happen first if it does play an important role in embrittlement, i.e., this is as logical as hydrogen having to get to a surface before it can bet into the enclosure of a crack or void.

In the event that the heat factor, Eq. 8-10, is large and exothermic (negative), the bulk term will be negligible, leaving only the surface term:

$$\phi = \frac{3}{2} N_{H'} \text{ (surface) } kT/A \qquad 8-11$$

and another explanation for embrittlement must be sought out for strong exothermic occluders.

OTHER THEORIES OF EMBRITTLEMENT

The evidence we have to go on is slight. Eq. 8-11 indicates that all metals of the strongly exothermic category would behave the same in the presence of internal hydrogen. These metals embrittle easily, as is known from powder technology. Powders of these metals are difficult to prepare because they are too soft, and hydrogen is usually introduced into these metals to make them brittle prior to grinding. Other theories, such as the decohesion theory (Petch, 1956), offer general treatments rather than specific for exothermic metals with the exclusion of the hydride formation theory. The metal-hydrogen bonds would naturally be more chemical-like in the instance of these metal "hydrides." It is, however, erroneous to try to explain embrittlement as being due to the formation of a chemical hydride because hydrogen is quite mobile in these systems. We mentioned earlier that the diffusion coefficient of hydrogen in titanium [ΔH of ~ -10 kcal (mole H)$^{-1}$] is 5.6×10^{-7} cm^2 sec^{-1}, or of about the same magnitude as that of hydrogen in palladium.

It is too much to expect that all these metals would have identical behavior with respect to hydrogen, but it seems likely that they are at least different theoretically from the endothermic occluders.

Possible insight lies in the w_{HH} term of Eq. 8-8 representing interactions between adsorbed protons. To be complete, this term should really be treated with the cluster theory. It is possibly of more significance than previously suspected.

This effect can be imagined as being important only in the vicinity of internal regions of high stress (Flitt & Bockris, 1981). There does not seem to be agreement as to whether the effect of w_{HH} will be repulsive or attractive.

In the exothermic occluders, concentrations are large, particularly near a region of high stress, and it seems reasonable to assume on this basis alone that w_{HH} would result in repulsive interactions between protons, allowing a large value in factor 8-9. To be of help, w_{HH} would have to have a positive value somewhere about 5 to 10 kcal mole^{-1}, and we will assume that this is the case. This possibility results in the same mechanism for embrittlement as before with Eq. 8-8, the bulk term (second) predominating and becoming large, for a different reason here. There is a hint of this large interaction energy already (Bockris et al., 1971) in the interaction energy of one mole of hydrogen into iron of 59 kcal.

Appendices

Appendix A
Infinite surface source, equilibrium, diffusing into a semi-infinite solid

Taking the case of infinite surface source, with equilibrium, diffusing into a semi-infinite solid, we have the following boundary conditions:

$$c(0, t) = c_o, \text{ constant, } t > 0 \qquad \text{A-1}$$

$$c(x, t) = 0, x \to \infty \qquad \text{A-2}$$

$$c(x, t) = 0, t \leq 0 \qquad \text{A-3}$$

where c is the concentration in moles cm^{-3}, x the distance into the semi-infinite solid, D the diffusion coefficient in $cm^2 \, sec^{-1}$, and c_o is the time constant equilibrium concentration just beneath the surface of the solid.

The transform solution is:

$$\bar{c} = ae^{-qx} + be^{qx} \qquad \text{A-4}$$

where:

$$q = (p/D)^{\frac{1}{2}} \qquad \text{A-5}$$

$$\bar{c} = c_o/p, \, x = 0 \qquad \text{A-6}$$

and

$$\bar{c} = 0, \, x \to \infty \qquad \text{A-7}$$

Proceeding to evaluate the constants in Eq. A-4

$$b = 0 \qquad \text{A-8}$$

and

$$\frac{c_o}{p} = a \qquad \text{A-9}$$

Thus,

$$\bar{c} = \frac{c_o}{p} e^{-qx} = \frac{c_o}{p} e^{-p^{1/2} x / D^{1/2}} \qquad \text{A-10}$$

and using the transform:

$$\frac{1}{p} e^{-kp^{1/2}} \leftrightarrow \operatorname{erfc} \frac{k}{2t^{1/2}} \qquad \text{A-11}$$

we have as a final solution:

$$\frac{c}{c_o} = \operatorname{erfc} \frac{x}{(4Dt)^{1/2}} \qquad \text{A-12}$$

Appendix B
Rediffusion from erfc initial condition

The initial conditions for the rediffusion from erfc initial condition start at t_o or $t = o$. Thus, we have:

$$c(x, o) = c_o \, \mathrm{erfc} \, \frac{x}{(4Dt_o)^{1/2}} \qquad \text{B-1}$$

and

$$c(o, t) = o \qquad \text{B-2}$$

That is, the concentration just beneath the surface for $t > o$ is zero.

We proceed, using separation of variables and eigen value expansion, assuming a product solution:

$$c = X(x) \, T(t) \qquad \text{B-3}$$

substituting into Fick's second law:

$$X''T = XT'/D \qquad \text{B-4}$$

dividing by XT:

$$X''/X = T'/DT \qquad \text{B-5}$$

Both sides may now be equated to a common constant, say μ.
For $\mu > o$:

$$T' = D \mu T \qquad \text{B-6}$$

$$T = (\text{constant}) \, e^{D\mu t} \qquad \text{B-7}$$

which can be rejected, since $T \to \infty$, with t.

$\mu = 0$ may also be rejected, and taking $\mu < 0$, say $\mu = -\lambda^2$, we get:

$$X'' = -\lambda^2 X \qquad \text{B-8}$$

$$X = a \cos \lambda x + b \sin \lambda x \qquad \text{B-9}$$

$$T' = -D \lambda^2 T \qquad \text{B-10}$$

$$T = (\text{constant}) \, e^{-D \lambda^2 t} \qquad \text{B-11}$$

and letting (constant) = 1:

$$c(x,t) = XT = [a \cos \lambda x + b \sin \lambda x] \, e^{-D\lambda^2 t} \qquad \text{B-12}$$

since $c(0, t) = 0 = a e^{-D\lambda^2 t}$: $\qquad \text{B-13}$

$$a = 0 \qquad \text{B-14}$$

$$c_\lambda (x, t) = b(\lambda) \, e^{-D\lambda^2 t} \sin \lambda x \qquad \text{B-15}$$

and integrating over λ to get the total or general solution:

$$c(x, t) = \int_{-\infty}^{\infty} b(\lambda) \, e^{-D\lambda^2 t} \sin \lambda x \, dx \qquad \text{B-16}$$

since:

$$c(x, 0)/c_0 = \text{erfc} \, \frac{x}{(4Dt_0)^{\frac{1}{2}}} \qquad \text{B-17}$$

we have the Fourier transform pair:

$$c_0 \, \text{erfc} \, \frac{x}{(4Dt_0)^{\frac{1}{2}}} = \int b(\lambda) \sin \lambda x \, dx \qquad \text{B-18}$$

$$b(\lambda) = \frac{1}{\pi} \left[\int_0^{\infty} c_0 \, \text{erfc} \, \frac{x}{(4Dt_0)^{\frac{1}{2}}} \right] \sin \lambda x \, dx \qquad \text{B-19}$$

with the integral:

$$\int_0^{\infty} \sin(2at) \, \text{erfc}(bt) \, dt = [1 - e^{-a^2/b^2}]/2a \qquad \text{B-20}$$

$$a'' > 0 \qquad \text{B-21}$$

APPENDIX B

$b'' > 0$ B-22

we get:

$$b(\lambda) = c_o [1 - e^{-(\lambda^2 D t_o)}]/\lambda \pi \quad \text{B-23}$$

$$c(x, t) = \left(\frac{c_o}{\pi}\right) \int_{-\infty}^{\infty} [1-e^{-\lambda^2 D t_o}] e^{-\lambda^2 D t} \sin \lambda x \frac{d\lambda}{x} \quad \text{B-24}$$

we get for the flux across $x = 0$:

$$J_t = -D (\partial c/\partial x)_{x=0} \quad \text{B-25}$$

$$J_t = c \left(\frac{D}{\pi}\right)^{1/2} [t^{-1/2} - (t + t_o)^{-1/2}] \quad \text{B-26}$$

Going back for $c(x, t)$, we use the integral:

$$\int_{-\infty}^{\infty} x^{-1} \sin a x \, dx = \pi \, \text{erf} \, (\tfrac{1}{2} a b^{-1/2}) \quad \text{B-27}$$

and get:

$$c(x, t) = c_o \left[\text{erf} \frac{x}{(2 Dt)^{1/2}} - \text{erf} \frac{x}{[4 D(t_o + t)]^{1/2}}\right] \quad \text{B-28}$$

Appendix C
Constant flux, diffusing into a semi-infinite solid

Constant flux diffusing into a semi-infinite solid, the initial and boundary conditions being:

$$c = 0, \quad t \leq 0 \quad \text{C-1}$$

$$-D \left(\frac{\partial c}{\partial x}\right)_{x=0} = j, \text{ constant}, \quad t > 0 \quad \text{C-2}$$

$$c = 0, \quad x \to \infty, \quad t > 0 \quad \text{C-3}$$

With the Laplace transformation, we get:

$$\bar{c} = ae^{-qx} + be^{qx} \quad \text{C-4}$$

$$-D \left(\frac{\partial \bar{c}}{\partial x}\right)_{x=0} = \frac{j}{p} \quad \text{C-5}$$

$$\bar{c}(x,t) = 0, \quad x \to \infty \quad \text{C-6}$$

From Eqs. C-4 and C-5:

$$\frac{j}{p} = -D(-aq + bq) \quad \text{C-7}$$

and from Eqs. C-4 and C-6:

$$b = 0 \quad \text{C-8}$$

Thus, from Eqs. C-8 and C-9:

$$a = \frac{-j}{pqD} \quad \text{C-9}$$

$$\bar{c} = \frac{j}{p^{3/2} D^{1/2}} e^{-p^{1/2} x / D^{1/2}} \quad \text{C-10}$$

APPENDIX C

and using the transform:

$$\frac{1}{p^{3/2}} e^{-kp^{1/2}} \leftrightarrow 2 \left(\frac{t}{\pi}\right)^{1/2} e^{-k^2/4t} - k \text{ erfc } \frac{k}{2t^{1/2}} \qquad \text{C-11}$$

or:

$$\leftrightarrow 2 t^{1/2} \text{ ierfc } k/2t^{1/2} \qquad \text{C-12}$$

we get:

$$c(x,t) = 2j \frac{t^{1/2}}{(\pi D)^{1/2}} e^{-x^2/4Dt} - \frac{jx}{D} \text{ erfc } \frac{x}{(4Dt)^{1/2}} \qquad \text{C-13}$$

or since:

$$c_o = c(o,t) = 2 j \left(\frac{t}{\pi D}\right)^{1/2} \qquad \text{C-14}$$

$$\frac{c(x,t)}{c_o} = e^{-x^2/4Dt} - \frac{x}{2} \left(\frac{\pi}{Dt}\right)^{1/2} \text{ erfc } \frac{x}{(4Dt)^{1/2}} \qquad \text{C-15}$$

References

CHAPTER 1

Bockris, J. O'M. 1972. Science 176:1323.

Daynes, H. 1920. Proc. Roy. Soc. 97A, 286.

Deluccia, J. J., and Berman, D. A. 1981. Electrochem. Corros. Testing, ASTM 727, Am. Soc. for Testing and Materials, 256.

Smith, D.P. 1948. Hydrogen in Metals. Chicago: University of Chicago Press.

CHAPTER 2

Abramowitz, M., and Stegun, I.A. 1965. Handbook of Mathematical Functions. New York: Dover.

Beck, W., Bockris, J. O'M., McBreen, J., and Nanis, L. 1966. Proc. Roy. Soc. A290:220.

Bockris, J. O'M., Beck, W., Genshaw, M. A., Subramanyan, P. K., and Williams, F. S. 1971. Acta Met. 19:1209.

Carslaw, H S., and Jaeger, J. C. 1959. Conduction of Heat in Solids. London: Oxford.

Crank, J. 1956. The Mathematics of Diffusion. London: Oxford.

Devanathan, M. A. F., and Stachurski, Z. 1962. Proc. Roy. Soc. A270:90.

Devanathan, M. A. F., Stachurski, Z., and Beck, W. 1963. J. Electrochem. Soc. 110:886.

Early, J. G. 1978. Acta Met. 26: 1215.

Flanagan, T. B., and Oates, W. A. Berichte der Bunsen - Gesellschaft. 1972. 76:706.

Frumkin, A. N., and Aladjalova, N. 1944. Acta Physicochim. URSS. 19:1.

Fullenwider, M. A. 1975. J. Electrochem. Soc. 122:648.

Gileadi, E., Fullenwider, M. A., and Bockris, J. O'M. 1966. J. Electrochem. Soc. 113:926.

McBreen, J., Nanis, L., and Beck, W. 1966. J. Electrochem. Soc. 113:1218.

Namboodhiri, T. K. G., and Nanis, L. 1973. Acta Met. 21:663.

Pressouyre, G. M., and Bernstein, I. M. 1978. Met. Trans. 9A:1571.

Schmidt, E., and Siegenthaler, H. 1970. Helv. Chim. Acta. 53:321.

Wach, S. 1971. Br. Corros. J. 6:114.

CHAPTER 3

Berman, D.A., Beck, W., and DeLuccia, J. J. 1974. Hydrogen in Metals, eds. I.M. Bernstein and A. W. Thompson. American Society for Metals.

Bockris, J. O'M., Genshaw, M. A., and Fullenwider, M. 1970. Electrochim. Acta. 15:47.

Bockris, J. O'M., Genshaw, M., and Subramanyan, P. K. 1967. Second Quarterly Report, Contract No. N00156-67-C-1941, Naval Air Eng.

Bockris, J. O'M., and Kita, H. 1961. J. Electrochem. Soc. 108:676.

Bockris, J. O'M., and Koch, D. F. A. 1961. J. Phys. Chem. 65:1941.

Bockris, J. O'M., and Subramanyan, P. K. 1971. Electrochim. Acta. 16, 2169.

Breiter, M. 1961. Electrochim. Acta. 6:25.

Carslaw, H. S. and Jaeger, J. C. 1959. Conduction of Heat in Solids. London: Oxford.

REFERENCES

DeLuccia, J. J., and Berman, D. A. 1981. Electrochem. Corros. Testing, ASTM 727, Am. Soc. for Testing and Materials, 256.

Devanathan, M. A. F., Bockris, J. O'M., and Mehl, W. 1959. J. Electroanal. Chem. 1:143.

Early, J. G. 1978. Acta Met. 26:1215.

Fullenwider, M. A. 1974. J. Electrochem. Soc. 121:313.

_____. 1975. J. Electrochem. Soc. 122:648.

_____. 1976. J. Electrochem. Soc. 123:197.

Johnson, H. H. 1967. Proceedings of Conference, Fundamental Aspects of Stress Corrosion Cracking, Columbus, p. 446, Nat. Assn. Corrosion Engineers (1969). Houston, Texas.

Miller, R. C., and Smits, F. M. 1957. Phys. Rev. 107:65.

Moelwyn-Hughes, E. A. 1964. Physical Chemistry. London: Pergamon Press.

Namboodhiri, T. K. G., and Nanis, L. 1973. Acta Met. 21:663.

Trapnell, W. H. 1955. Chemisorption. London: Butterworths.

CHAPTER 4

Bockris, J. O'M., Genshaw, M. A., and Fullenwider, M. A. 1970. Electrochim. Acta. 15:47.

DeLuccia, J. J. 1975. Thesis, U. of Pa.

Fowler, R., and Guggenheim, E. A. 1965. Statistical thermodynamics, London: Cambridge.

Fullenwider, M. A., 1974. J. Electrochem. Soc. 121:1589.

Lacher, J. R. 1937. Proc. Roy. Soc. A161:525.

Thompson, A. W., and Bernstein, I. M., eds. 1975. Effect of Hydrogen on Behavior of Materials. Proceedings of an International Conference, Jackson Lake Lodge, Moran, Wyoming, Sept. 7-11, 1975, Pub. of the Met. Soc. of AIME.

CHAPTER 5

Bockris, J. O'M., and Reddy, A. 1970. Modern electrochemistry. New York: Plenum.

Fullenwider, M. A. 1974. J. Electrochem. Soc. 121:1589.

Gileadi, E., ed. 1967. Electrosorption. New York: Plenum.

Gileadi, E., Fullenwider, M. A., and Bockris, J. O'M. 1966. J. Electrochem. Soc. 113:926.

McBreen, J., and Genshaw, M. A. 1967. Proc. Conf. Stress Corrosion Cracking, Ohio State U., Columbus, Ohio, September 1967, Pub. Nat. Assoc. Corros. Eng., Houston, Texas.

Richardson, O. W., Nicol, J., and Parnell, T. 1904. Phil. Mag. (VI) 8:1.

CHAPTER 6

Bocciarelli, C. V. 1967. Solid State Comm. 5:821.

Bockris, J. O'M., Genshaw, M. A., and Fullenwider, M. A. 1972. Electrochim. Acta. 15:706.

Boehm, H., Treptow, W., Wunsch, G., Kiener, V., Meger, H., and Csizi, G. 1979. U. S. Patent #4,153,742.

Bryan, W. L., and Dodge, B. F. 1963. J. Am. Inst. Chem. Engrs. 9:223.

Ebisuzaki, Y., Kass, W. J., and O'Keefe, M. 1967. J. Chem. Phys. 46:1378.

Flanagan, T. B., and Oates, W. A. 1972. Berichte der Bunsen - Gesellschaft. 76:706.

Fullenwider, M. A. 1977. U. S. Patent #4,031,291.

Ham, W. R. 1932. J. Chem. Phys. 1:476.

Handbook of Chemistry and Physics 1974. 55th ed. Chemical Rubber Co. Cleveland, Ohio.

Heikel, H. R., and Leddy, J. J. 1980. U. S. Patent #4,214,971.

Leaman, F. H. 1972. U. S. Patents #3,153,742 and #3,698,939.

Maeland, A. J., and Gibb, T. R. P., Jr. 1961. J. Phys. Chem. 65:1270.

McBreen, J., Nanis, L., and Beck, W. 1966. J. Electrochem. Soc. 113:1218.

Richardson, O. W., Nicol, J. and Parnell, T. 1904. Phil. Mag. 8:1.

Schuldiner, S., and Warner, T. B. 1965. J. Electrochem. Soc. 112:212.

REFERENCES

Spaziante, P. M., and Nidola, A. 1980. U. S. Patent #4,214,970.

CHAPTER 7

Flanagan, T. B., and Oates, W. A. 1972. Berichte der Bunsen - Gesellschaft. 76; 706.

Johnson, J. R., and Reilly, J. J. 1978. Inorg. Chem. 17:3103.

Reilly, J. J. 1978. Proc. Int. Symp. on Hydrides for Energy Storage, Geilo, Norway, 1977, Ed. A. F. Andresen and A. Milland. London: Pergamon.

_____. 1979. Z. Phys. Chem. Neue Folge. 117:655.

Reilly, J. J., and Wiswall, R. H., Jr. 1974. Inorg. Chem. 13:218.

Van Mal, H. H. 1976. Stability of Ternary Hydrides and Some Applications, Thesis, Tech. Hogeshool, Delft.

Van Mal, H. H., Bushow, K. H. J., and Miedema, A. R. 1975. J. Less Common Metals. 35:283.

Wagner, C. 1971. Acta Met. 19:843.

Yamanaka, K. Saito, H., and Someno, M. 1975. Nippan Kagaku Kaishi (J. Chem. Soc., Japan). 8:1256.

CHAPTER 8

Beck, W., Bockris, J. O'M., McBreen, J., and Nanis, L. 1966. Proc. Roy. Soc. A290:220.

Benneck, H., Schenck, H., and Muller, H. 1935. Stahl and Eisen. 55:321.

Bockris, J. O'M., Beck, W., Genshaw, M. A., Subramanyan, P. K., and Williams, F. S. 1971. Acta Met. 19:1209.

DeLuccia, J. J., and Berman, D. A. 1981. Electrochem. Corros. Testing ASTM STP 727 Florian Mansfeld and Ugo Bertocci, eds., American Soc., A290:220.

Flanagan, T. B., and Oates, W. A. 1972. Berichte der Bunsen - Gesellschaft. 76:706.

Flitt, H. J., and Bockris, J. O'M. 1981. J. Hydrogen Energy. 6:119.

Gileadi, E., Fullenwider, M. A., and Bockris, J. O'M. 1966. J. Electrochem. Soc. 113:926.

Gosar, P. 1964. Nuovo Cimento. 31:781.

Griffith, A. A. 1929. Phil. Trans. A221:163.

Ham, W. R. 1932. J. Chem. Phys. 1:476.

Oriani, R. A. 1972. Ber. Bunsen. Ges. Phys. Chem. 76:848.

Petch, N. J. 1956. Phil. Mag. 1:331.

Sieverts, A. 1907. Z. Phys. Chem. 60:169.

_____. 1910. Z. Electrochem. 16:707.

_____. 1920. Z. Metallk. 21:37.

_____. 1930. Metallurgist. 5:168.

Subramanyan, P. K. 1970. Thesis, U. of Pa.

Troiano, A. R. 1960. Trans. A. S. M. 52:54.

Index

Absolute activity, 54
Absolute surface coverage of palladium with hydrogen, 15, 23
Adsorption isotherms, 65, 66
Analysis for the barnacle electrode, 28, 32, 40
Anomalies in permeation behavior 12, 38

Barnacle electrode, 26, 95
Bi-electrode gasket, 6, 14, 17
Bi-electrode technique, 5

Cadmium electroplating solutions, 11
Catalysis, 2, 78
Critical current density, 99
Constant flux concept, 16, 22, 35, 39
Coupled discharge combination, 47
  electrochemical, Langmuir, 50-69
  electrochemical, Tempkin, 72

deFord type circuit, 5
Diffusion coefficient for hydrogen in metals where surface effects play no part, 17
Diffusion equation, 4

Energy change upon introducing one mole of H into iron, 14
Embrittlement, 2, 41, 95
Epitaxial powders, 84, 94
Estimation of hydrogen concentration in a diffused specimen, 27
Exchange current density, 64
External hydrogen, 53, 55, 57, 58, 61, 62, 81

Fast discharge
  slow electrochemical, Tempkin, 47-71
  slow recombination, Langmuir, 45-67
Fick's second law, 4
Fuel cells, 2, 78, 85

Gibbs free energy of adsorption of a species, 59

Grand partition functions, 53-61
Griffith crack, 100

Heat transfer, 4
Hydrogen diffusion as a function of compressive stress, 13
Hydrogen diffusion
  as a function of dissolved hydrogen concentration, 12
  as a function of temperature, 12, 13
  as a function of tensile stress, 12, 13
  in iron, 5, 12, 23
  in palladium, 5, 10, 14, 23
  in single crystal iron, 12
  in 4340 steel, 11, 13
  in zone refined iron, 12
Hydrogen economy, 2, 88
Hydrogen oxidizing catalysts, 78

"Impenetrable surface" boundary condition, 28
Input control for hydrogen entry, 17, 18, 23
Internal hydrogen, 53, 59, 60, 61, 62, 81, 103, 104
Iron - titanium alloys, 93

Linear concentration profile, 20

Mechanism of catalytic action of platinum, 84

Oxygen reducing catalysts, 78

Partial molar volume, 13
Palladium plating solution, 11

Phonon flux, 52
Possibilities for getting hydrogen into the metal, 39, 40, 41

Rate constants for the hydrogen reaction on palladium, 73
Rate limitation
  permanent "skin," 35, 36, 82
  removable "skin," 38, 82, 102
Rediffusion, 28, 33, 35
Rules followed by ternary alloys, 88
Rule of reversed stability, 89

Slow discharge
  fast electrochemical, Langmuir, 46-68
  fast recombination, Langmuir, 45-66
  fast electrochemical, Tempkin, 70
  fast recombination, Tempkin, 72
Spreading pressure for internal and external hydrogen, 77
Structures of alloys, 91

Temperature, 53
Thermodynamics of hydrogen electrode, 75
Thermodynamics of metal hydrogen systems, 1
Thermal equilibrium concept, 8, 22, 35, 40
Three dimensional, translational, ideal gas partition function, 55
Total grand partition function factor for metal-hydrogen systems, 55

INDEX

Two dimensional translational, ideal gas partition function, 59

Variation of the diffusion coefficient with concentration, 23, 38

About the Author

MALCOLM FULLENWIDER is a graduate of Albion College, Michigan, and the University of Pennsylvania, Philadelphia. He is currently the president of a small consulting company, RK Laboratory, Whitehall, Pennsylvania.

The author has a wide background including taxicab driving in Detroit, Michigan; alcohol-drug abuse counselling in Dayton, Ohio; working as a security guard in Allentown, Pennsylvania; a brief experience as an industrial chemist; and a post doctorate in Bern, Switzerland.

His scientific interests range from electrochemistry, hydrogen in metals, the hydrogen economy, crystallography, the mathematics of diffusion, to the time variation of the fundamental constants.

Dr. Fullenwider is a member of the American Chemical Society, the American Institute of Physics, the Electrochemical Society and is an associate member of the International Association for Hydrogen Energy.

Sidney Howard and Clare Eames